THE DYNAMIC EARTH

A SCIENTIFIC AMERICAN Book

THE DYNAMIC EARTH

W. H. FREEMAN AND COMPANY
New York San Francisco

The Cover

The computer-graphics image on the cover symbolizes the theme of this issue of SCIENTIFIC AMERICAN: the dynamic earth. The image shows an aurora over the earth's north magnetic pole viewed from a satellite at an altitude of 3.19 earth radii. The aurora is the luminous ellipse projected on the green computer-generated map of the globe. The bright crescent at the upper left is the illuminated daylight side of the earth. Auroras are emitted by molecules in the earth's upper atmosphere excited by electrically charged particles in the "wind" of thin gas expanding outward from the sun. The particles plunge into the atmosphere because they are trapped in the earth's magnetic field, which is generated by slow flows in the earth's metallic core. Therefore the aurora is a manifestation of the dynamic activity of the core. The image was made by a University of Iowa team from data transmitted by instrumentation aboard the satellite *Dynamics Explorer 1.* It is provided through the courtesy of Louis A. Frank.

Library of Congress Cataloging in Publication Data

Main entry under title:

The Dynamic earth.

Published also as v. 249, no. 3 of Scientific American.
"A Scientific American book."
Bibliography: p.
Includes index.
1. Geodynamics—Addresses, essays, lectures.
QE501.D96 1983 551 83-20493
ISBN 0-7167-1611-9

The eight chapters in this book originally appeared as articles in the September 1983 issue of *Scientific American.*

Printed in the United States of America

1234567890 DO 1089876543

CONTENTS

FOREWORD

What makes the Earth so interesting—and our presence on it possible—is that it is a dynamic planet. Immense forces, turning cyclically in the core, in the surrounding mantle, in the crust, and in the air, oceans, and biosphere in which we live, keep the Earth in a steady state of constant change.

As everyone knows by now, the continent under our feet is in motion. North America drifts westward about one inch a year. That is slow motion, but not imperçeptible; it can be observed in the growth of a fingernail. The recognition of continental drift two decades ago [see Continental Drift by J. Tuzo Wilson, SCIENTIFIC AMERICAN, April 1963] brought a revolution in geology. A first comprehensive statement of the consequent revision of the picture of our planet is undertaken by the authors of this book, each a contributor to a worldwide collaborative work that remains very much in progress.

In this new picture of Earth, the ocean floor holds the foreground along with the familiar dry land of the continents. Forces originating in the mantle and in the core at the center are seen to play as large a role in shaping and changing the world map as the more visible cycles of hydrosphere, atmosphere and biosphere. In interlocking exchange with these cycles, the crustal lithosphere turns over in a cycle that regenerates the eroding continents from new sea floor rock.

This system has an ongoing history. The sorting out of the elements between core and surface by differentials in pressure, temperature and chemical affinity liberated Earth's water and air from the rock and, over the first aeons, floated the lighter rock of the continents in the crust. Life became a geological force when single-celled organisms first acquired the metabolic capacities of photosynthesis and oxygen respiration, establishing and maintaining the atmosphere in its present elemental composition.

New crustal rock continues to be generated by upwelling of mantle material under the rifted ridge that runs 40,000 miles around the Earth on the ocean floor, a feature of the world map as grand as the continents. Spreading outward from the ridge, the new ocean floor pushes the continental rafts this way and that. Where there are deep trenches at the margins of the continents, the sea floor plunges underneath, carrying with it material outwashed from the continents and the oozes loaded with detritus of ocean life, to form new continental rock in magma chambers inshore under the continental crust. From these chambers rise the great cordilleran mountain chains.

Thus, as the anatomy and physiology of the system become better understood, the features of the world around us find a coherent, unified explanation. Francis Bacon's famous hunch—he perceived the jigsaw fit of the shorelines of South America and Africa on one of the first maps ever to show those shorelines—stands confirmed: a great earlier continent was indeed split apart by the opening up of the Atlantic. The occurrence of the same fossil "glossopteris" leaves in the coals of India, Australia, South America, South Africa and Antarctica testifies to their union 500,000 millennia ago in the continent called Gondwanaland. The "ring of fire" around the Pacific basin is plainly generated by crustal subduction. The heaping up of the lofty Himalayas can be seen as the consequence of the collision of the Indian continental plate with the Eurasian.

Because scientific knowledge is, in Benjamin Franklin's phrase, "useful knowledge," the new geology makes its contribution to technology. Leaching of the up-welling mantle rock by seawater lays down concentrations of nonferrous metals along the submarine rifted ridge; prospectors now direct their treasure hunt to fossil ridges ashore. Similarly, the continental shelves are recognized as major traps for petroleum and natural gas, directing prospectors to look again for fossil shelves on continents.

Strengthened by knowledge, man himself must now be reckoned as a geological force. The steadily rising concentration of carbon dioxide in the atmosphere caused by the burning of fossil fuels portends climatic changes of a kind and on a scale not yet securely understood or predictable. "Large as the earth is," says Raymond Siever in the first, introductory chapter, "the behavior of one of its biological species may be enough to seriously disturb its balance."

The chapters of this book first appeared in the September 1983 issue of SCIENTIFIC AMERICAN, the thirty-fourth of the annual single-topic issues published by the magazine. It has been made available in book form by the expeditious work of W. H. Freeman and Company, the book-publishing affiliate of SCIENTIFIC AMERICAN.

THE EDITORS*

September 1983

1

THE DYNAMIC EARTH

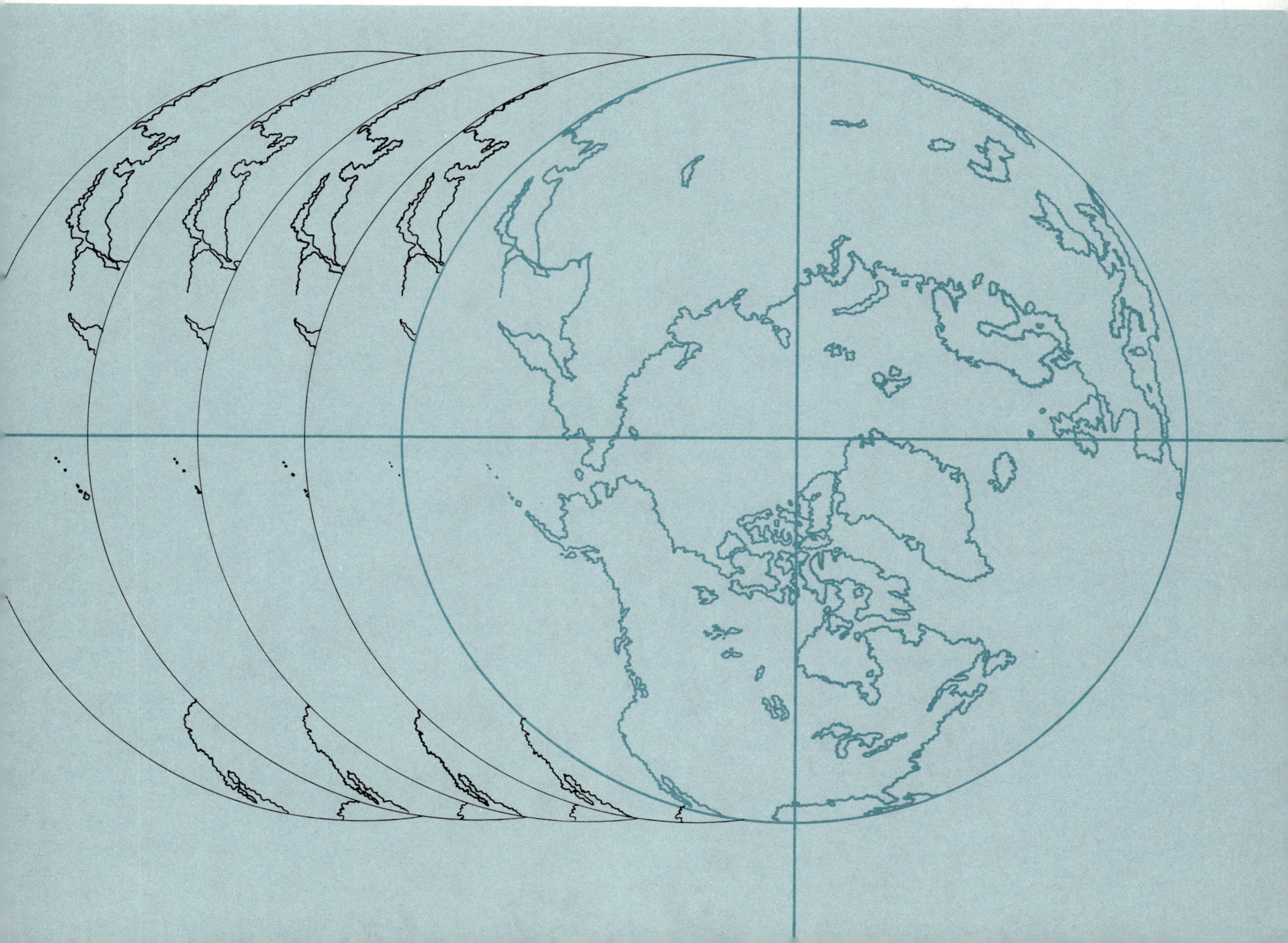

The Dynamic Earth

by RAYMOND SIEVER

Presenting a volume on the earth as a system of interacting fluids, including living matter. Some of the flows are fast, others slow, but overall the planet is maintained in a remarkably steady state

Scientists who study the earth are accustomed to working with many different scales of space and time. The physical dimensions of their subject range from the global to the submicroscopic, from volumes of matter measured in cubic kilometers to interatomic gaps measured in angstrom units. Often a given research topic will encompass widely disparate scales of length, as when an earthquake triggered by a slippage of a few centimeters along a fault generates seismic waves that travel for thousands of kilometers through the earth's interior. Similarly, the temporal dimensions of geology include not only short-lived phenomena such as earthquakes, volcanic eruptions and meteorite impacts but also events recorded in units of tens or hundreds of years (for example the meandering of a river), thousands of years (glaciation), millions of years (continental drift) and even billions of years (the formation of the present, oxygen-rich atmosphere). Again a single process, notably weathering, can be studied over a wide range of time scales: from the minutes and hours of a laboratory experiment measuring the dissolution rate of a mineral to the thousands of years needed to form a soil. Taken in various combinations, the parameters of geologic space and time define the scope of this single-topic volume of *Scientific American:* the multitude of great and small changes that have taken place—and continue to take place—in the history of the earth.

Most geologists, oceanographers and other earth scientists tend at one time or another to think of the earth as a machine, or perhaps as a living organism. The metaphor of the machine captures an important aspect of the earth's dynamism: in spite of all the changes that are observed at many different scales of space and time, the earth as a whole stays remarkably constant. In particular, it has become apparent in recent years that major parts of the globe—the core, the mantle, the crust, the oceans and the atmosphere—can be profitably viewed as a complex, interacting system in which there is a cyclic flow of material from one reservoir to another. The mechanistic model of the earth as a vast recycling system has its counterpart in the physiological model of dynamic equilibrium known as homeostasis.

The hierarchy of scales that pervades the work of the earth scientist is perhaps best exemplified in the making of a geologic map, an artifact that can be described in somewhat nongeologic terms as a graph of the position on a coordinate system at the earth's surface of rock formations of different ages. The first step in geologic mapping is the job of the field investigator, who determines two major properties of the rocks at a given site: age and composition. At a typical rock outcrop one can see only small-scale relations, usually over distances measured in meters. From an assemblage of such observations the final map of the region is constructed, like any other graph, by interpolation and extrapolation, showing forms appropriate to its scale. On a map with an area of, say, 200 square kilometers one can see patterns of river valleys and the characteristic folds and faults of bedrock. The wealth of information at the outcrop level is sacrificed in favor of grosser features. On a map that covers a region of many thousands of square kilometers one begins to see even larger features: plateaus, mountains, plains, entire river systems, the outlines of a rift valley, the distribution of glacial lakes. As the map gets to continental or global proportions one sees the largest structures of the continental surface, principally mountain chains. Whenever detailed information is sacrificed for patterns that show up only on a large-scale map, the trick is to recognize what details to leave out. In other words, the essence of this kind of geologic analysis is always to separate the "signal" from the "noise."

Inevitably earth scientists face the problem of reconciling scales. For example, structural geologists and geophysicists are currently trying to relate the large-scale collisions of tectonic plates that throw up high mountains such as the Alps and the Himalayas to the small-scale folds and faults that can be seen on an individual mountainside. The aim is to learn how to go in the opposite direction: to infer from small-

EVIDENCE OF ANCIENT UPHEAVALS underlies the pastoral scene on the opposite page. The engraving is from James Hutton's *Theory of the Earth,* published in 1795. It shows an exposed riverbank of the river Jed in southern Scotland. The vertical beds of rock at the bottom of the bank were originally laid down as oceanic sediments. They subsequently underwent metamorphosis to become schist and were deformed and uplifted to become part of a mountain chain. The band of mixed material just above them is erosional debris dating from that period. The metamorphic rocks were then submerged again, and the horizontal sedimentary beds of sandstone were deposited above them. Finally the entire formation rose above sea level once more and the new soil layers were formed at the top. Hutton cited such examples from his wide-ranging field trips as evidence both of the earth's antiquity and its dynamic activity. In modern terminology a rock formation of this type is called an angular unconformity.

scale folds and faults what an older, obliterated mountain chain was like and how it might have been created by ancient plate motions.

Scales of time involve additional subtleties. A river flowing at several centimeters per second moves at roughly the same speed as a fault block slipping in an earthquake. Yet the distribution of these two movements over a longer period is quite different. The river runs more or less steadily for months or years, whereas the movements along the fault are episodic, with intervals of little or no movement extending over as much as hundreds of years. The distinction between steady and episodic change is central to the current debate over the rate of biological evolution: the key question here is whether or not the geologic time scale embodied in fossil-bearing rocks is precise enough to resolve the issue between the "gradualist" and the "punctuated equilibrium" models of the evolution of species.

Generations of geologists have relied primarily on a biological clock: the sequences of fossils that by their evolutionary changes mark the major divisions of geologic history. In the 19th century that was the only choice; in the 20th century, however, it became possible to calibrate the fossil clock by another timekeeper: the radioactive-decay clock, which depends on the known decay rates of radioactive isotopes of carbon, uranium, potassium, rubidium and neodymium. The scales of the events the two clocks date are apt to be different, as is the nature of the events timed. Working with both clocks is a little like arranging to meet someone at a certain time when one person relies on his pulse to count seconds and the other has a watch with only an hour hand.

Geologists refer to the apparent motion of the sun (or what might be called everyday time) to time comparatively fast processes such as the weather, floods, landslides, volcanic eruptions and earthquakes. They invoke radioactivity alone to time very slow processes such as the evolution of the atmosphere. In between lies the scale of geologic time over which continents move, mountain chains rise, the earth's magnetic field reverses, fossil species evolve

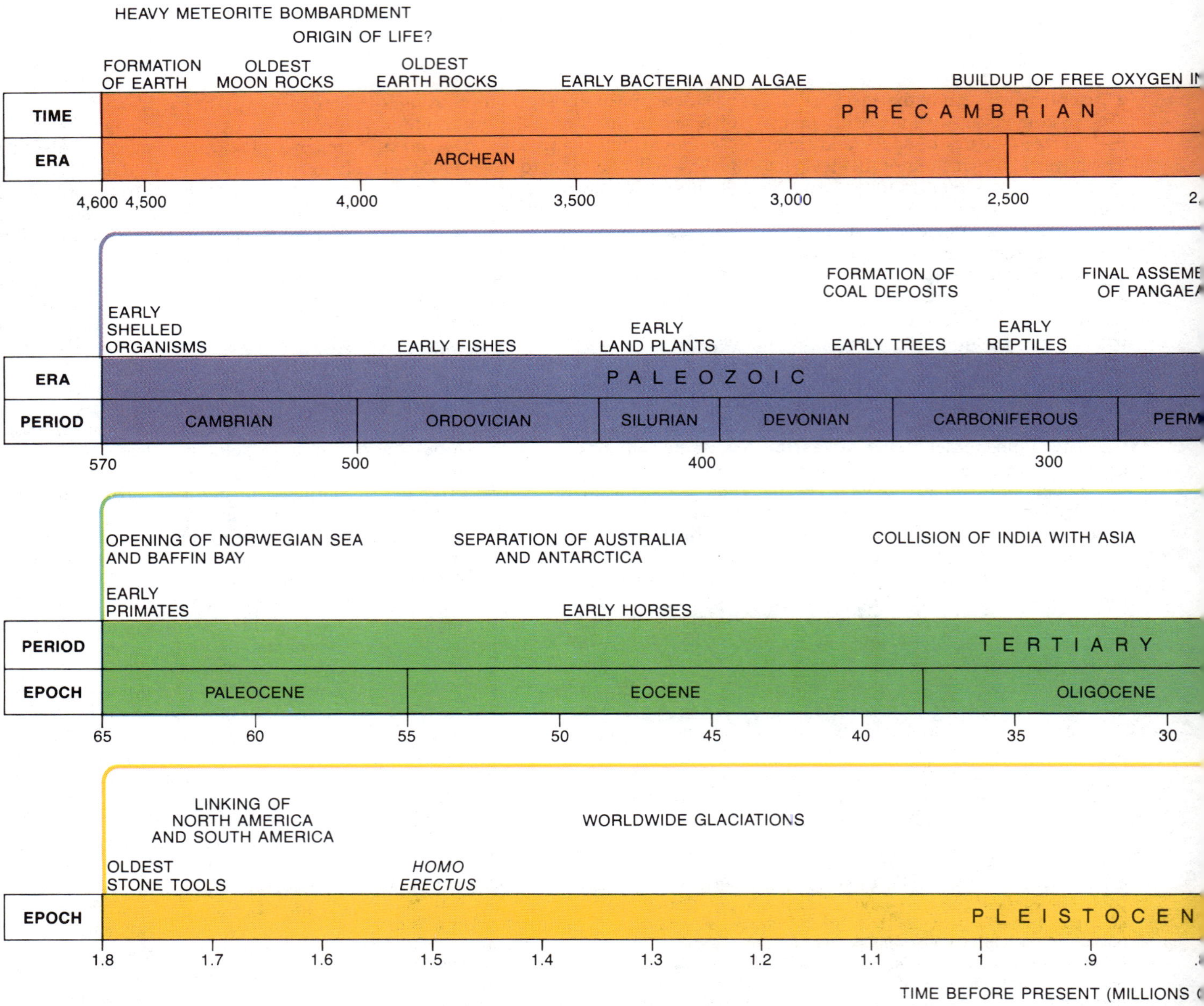

GEOLOGIC TIME SCALE, originally constructed by 19th-century naturalists solely on the basis of fossil evidence, has been calibrated by means of modern radioactive-dating techniques. In this representation the top line shows the full sweep of geologic time from the origin of the earth some 4.6 billion years ago to the present. The comparatively short span of Phanerozoic time, during which fossils of shelled organisms have been abundant in the geologic record, is enlarged in the second line from the top, and successively shorter time segments are enlarged in the next two lines. The three eras of Phanerozoic time (Paleozoic, Mesozoic and Cenozoic) are divided in turn

and glacial epochs last. Over this intermediate time scale reference is made primarily to the specific rock sequences that constitute the geologic record.

For example, in order to study the meandering of a river one can rely on historical records if they go back far enough, perhaps supplementing this information with data drawn from the relicts of prehistoric terraces. To study the long-term evolution of the river, from the early cutting of its valley into bedrock to its later broadening into a characteristic floodplain, however, one has no alternative but to consult the geologic record. The rise in sea level since the retreat of the glaciers some 10,000 years ago is also revealed in the geologic record, which yields information both on the rate of change of the ice sheets and their relation to the sea and on an important property of the earth's interior. As the ice retreated it unloaded the crust, and from the crust's "rebound" it is possible to estimate the viscosity of the mantle material that flowed under it in order to accommodate the rise.

The history of the earth is studied not only for its own sake but also for economic reasons: to explore for oil, gas, metallic ores and other useful substances formed at a specific time and place in the distant past. Simple curiosity, however, is what drives most geologists to want to know just what happened when. They want to know about the last ice age both for what it can reveal about the next one and for a glimpse of what the conditions of life were for early man. Finally, there are those geologists who specialize in the search for evidence that the history of the earth is not just one random thing after another but rather the playing out on a long time scale of the cycles of a great machine.

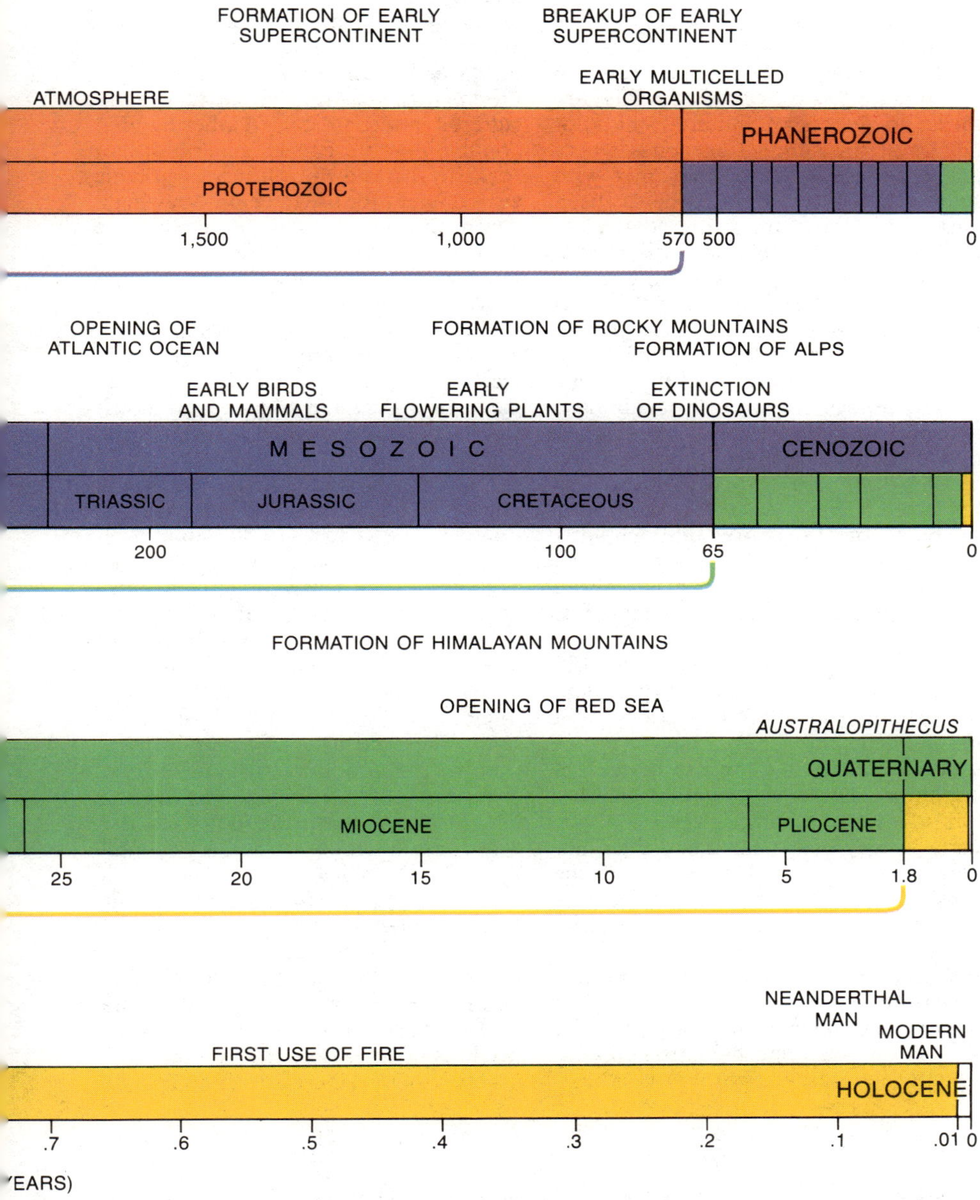

into 11 periods; the Tertiary period is divided into five epochs, and the Quaternary period consists of the Pleistocene and Holocene epochs. The calibration of the geologic clock by radioactive dating is a continuing concern. For example, according to a recent review of the radiometric evidence by an international group headed by the French geochronologist G. S. Odin, the beginning of the Cambrian period may have been between 540 and 520 million years ago.

If earth history is cyclic, over what periods of time does it repeat itself, and how does one keep track of the process? The hydrologic cycle serves as a model of the cyclic flow of material between different parts of the earth. Water from one major reservoir, the atmosphere, falls on the land and the sea. Some of the precipitation is stored temporarily in the ground and in lakes. The rest follows various routes back to the other major reservoir, the sea. Evaporation from the sea and the land back to the atmospheric reservoir completes the cycle.

The hydrologic cycle is global. Hence by adding up all the water in all the world's reservoirs and flow paths one can arrive at an estimate of the total water content of the system and its major parts. Bypassing analysis of individual reservoirs has the effect of removing heterogeneity: one finds that the global balance is quite steady from year to year. In other words, there is always about the same amount of water in the atmosphere, in the oceans, in the polar ice caps and on the continents. Over periods of less than a year the system may not be so steady, and over periods of a few years the global balance may shift somewhat from one reservoir to another. Indeed, perturbations of the system can elucidate the operation of the cycle. The ice ages represented one such shift in the cycle: water was removed from the ocean and stored in the ice reservoir, and the flows from one reservoir to another were adjusted accordingly; the result was a drastic change in the climate and a lowering of the mean sea level, which exposed much of the area of the continental shelves. The continuing discussion of this major perturbation centers on the possible reasons for the shift and how rapidly the glacial ice caps expanded and then shrank as the earth returned to the comparatively unglaciated stage that prevails today. Polar ice and mountain glaciers still exist, of course, and so it is not possible to know from firsthand evidence what the hydrologic cycle was like when the earth was completely free of ice.

The movement of carbon dioxide through the atmosphere, the ocean and the solid earth provides another opportunity for following the large-scale flow of chemical elements from one part of the earth to another. The carbon dioxide of the atmosphere is taken up by plants in photosynthesis and by rocks in weathering. Photosynthesis manufactures the earth's supply of organic carbon and weathering accounts for the calcium carbonate of limestones, a major product of the transformation of igneous rocks into sediments. Dead plant and animal matter, together with the calcium carbonate shells of mollusks and other organisms, immobilizes the carbon. As these materials are buried in the form of sediment and incorporated into the crust of the earth, carbon is removed from the surface reservoir. At the same time ancient organic matter and limestones are eroded and chemically weathered. By the oxidation of organic matter and the dissolution of calcium carbonate, carbon dioxide is returned to the dynamic system, keeping it in balance.

In much the same way one can construct cycles of all the elements and their isotopes. For example, from the rate at which dissolved calcium enters the ocean from rivers (about 10^{13} moles per year) and the total amount of calcium in the oceans (about 10^{19} moles) one can estimate how long an ion of calcium can be expected to remain in the ocean: about a million years. On the average that is how long it takes before the calcium combines with a carbonate group and settles out of the reservoir as a part of some limestone. The calcium ion may then be buried, perhaps ultimately to be incorporated into a silicate-bearing metamorphic rock. Alternatively, it may go deeper still, becoming part of a magma that will return to the surface in a congealed form as an igneous rock, there to be weathered, dissolved and sent back to the ocean by a river.

These chemical cycles are different versions of the grand geologic cycle that James Hutton, the founder of modern geology, enunciated almost 200 years ago. In Hutton's original version rocks are weathered to form sediment, which is then buried. After deep burial the rocks undergo metamorphosis and/or melting. Later they are deformed and uplifted into mountain chains, only to be weathered again and recycled. In spite of many arguments and theories about the mechanisms of Hutton's rock cycle its basic outlines remain intact as the geologist's way of viewing an earth of constant change.

The continental crust, the repository of the geologic record of the past 3.8 billion years, is engaged in its own cycle of destruction and renewal. Each year some 10^{16} grams of solid and dissolved products of the erosion of the land surface are removed by rivers, wind and glacial ice. Much of the solid detritus is deposited on the continental shelves, but a good deal is lost to the ocean basins. The return is by way of subduction zones, where some of the sediments of the deep-sea basins are scraped off as the oceanic lithosphere plunges into the mantle [see "The Oceanic Crust," by Jean Francheteau, page 39]. The oceanic sediment, which is found in front of oceanic island arcs and along continental margins where the subduction zone borders a continental mass, is plastered back onto the continent. The igneous rocks generated at subduction zones are also added to the continents. Thus over geologic time the continental masses remain in a steady state, even though there are frequent rises and falls in sea level that periodically flood and bare the continental shelves and the lower parts of the continents proper. The sediments and igneous rocks that are added to the continents are welded to them mainly as mountain chains associated with plate boundaries [see "The Continental Crust," by B. Clark Burchfiel, page 55]. Old mountains, then, are the remnants of a past recycling of the continents brought about by the motions of the lithospheric plates.

The recycling story extends deep into the mantle, as crustal materials are thrust to depths of hundreds of kilometers at the subduction zones where plates converge. There they mix with some materials that have never been part of the crust and others that have reached the surface before in their history. Geologists are just beginning to understand how the materials of the earth mix under the conditions of high temperature and pressure prevailing in the interior. The chief tracers of this mixing are the radioactive isotopes of rubidium and neodymium, which are helping to reveal the relations between age and mixing [see "The Earth's Mantle," by D. P. McKenzie, page 25]. Studies of the behavior of the core and the mantle over various time scales will determine to what extent they too can be viewed as parts of the great recycling machine.

If the machine has been going for billions of years, including some violent perturbations at times, how was it built in the first place? By what series of steady states did it evolve to its present condition? These are the questions that link a fragmentary rock record of the early history of the earth and the other planets to deductions drawn from the astronomy of the formation of the stars and the evolution of the solar system.

Theories of the formation of the solar system out of a nebula of gas and dust are still being refined, but they all have in common the central idea that about 4.6 billion years ago the earth grew to roughly its present size by a combination of two processes: the condensation of primitive solar-nebula material and the accretion of bits and pieces of other planetary matter in the vicinity. The early history of the earth was marked by continued accretion and by a rapid increase in temperature, due to a combination of three effects: heating from the radioactive elements abundant in the early condensate, heating from the impact of infalling material and heating from the contraction of the young planetary body. The rise in temperature, according to the now conventional view, led to widespread melting and a massive differentiation of the earth into a core, a mantle and a crust. All these ideas, proposed long before the exploration of the solar system by spacecraft, were refined in recent years by studies of the moon and of other planets. This was particularly true of the moon, where astronauts were able to sample a body whose history was frozen at an early stage. The moon, with no atmosphere or oceans, suffered no chemical weathering that could obliterate earlier generations of rock. It also did not offer a hospitable environment for life to begin. Viewing it emphasizes how vital the fluid envelope of water and gas was to the earth's machine. The composition of the gas, however, was not the same as that of the atmosphere of today. In its earliest stages the primitive atmosphere was free of oxygen and contained reduced gases such as methane and ammonia [see "The Atmosphere," by Andrew P. Ingersoll, page 81].

How the dynamics of the earth machine worked in the early days is a deductive study, because no mountains or sediments are preserved to give any hint of the products. The radioactively dated rock record begins in a fragmentary way about 3.8 billion years ago, which is the age of metamorphism by pressure and temperature of a series of originally sedimentary rocks in southwestern Greenland. The record clearly demonstrates that the essential operations of the geologic machine were similar to those in effect today. Iron formations and other types of sedimentary rocks such as sandstones and shales can be recognized in their metamorphosed forms in the Greenland rocks. Igneous rocks found at the site seem to have been created by much the same kind of melting that can be seen operating today. The deformation of rocks is also similar to the deformations of later times.

Nevertheless, there are differences be-

CHANGING FACE OF THE EARTH is portrayed at 60-million-year intervals from the Cambrian period to the present in the 10 computer-generated maps on the opposite page. The sequence documents the assembly and breakup of the supercontinent of Pangaea, the two chief episodes of continental drift leading up to the present arrangement of the earth's land masses. The maps were assembled by Alfred M. Ziegler and Christopher S. Scotese of the University of Chicago's Paleographic Atlas Project mainly on the basis of paleomagnetic evidence.

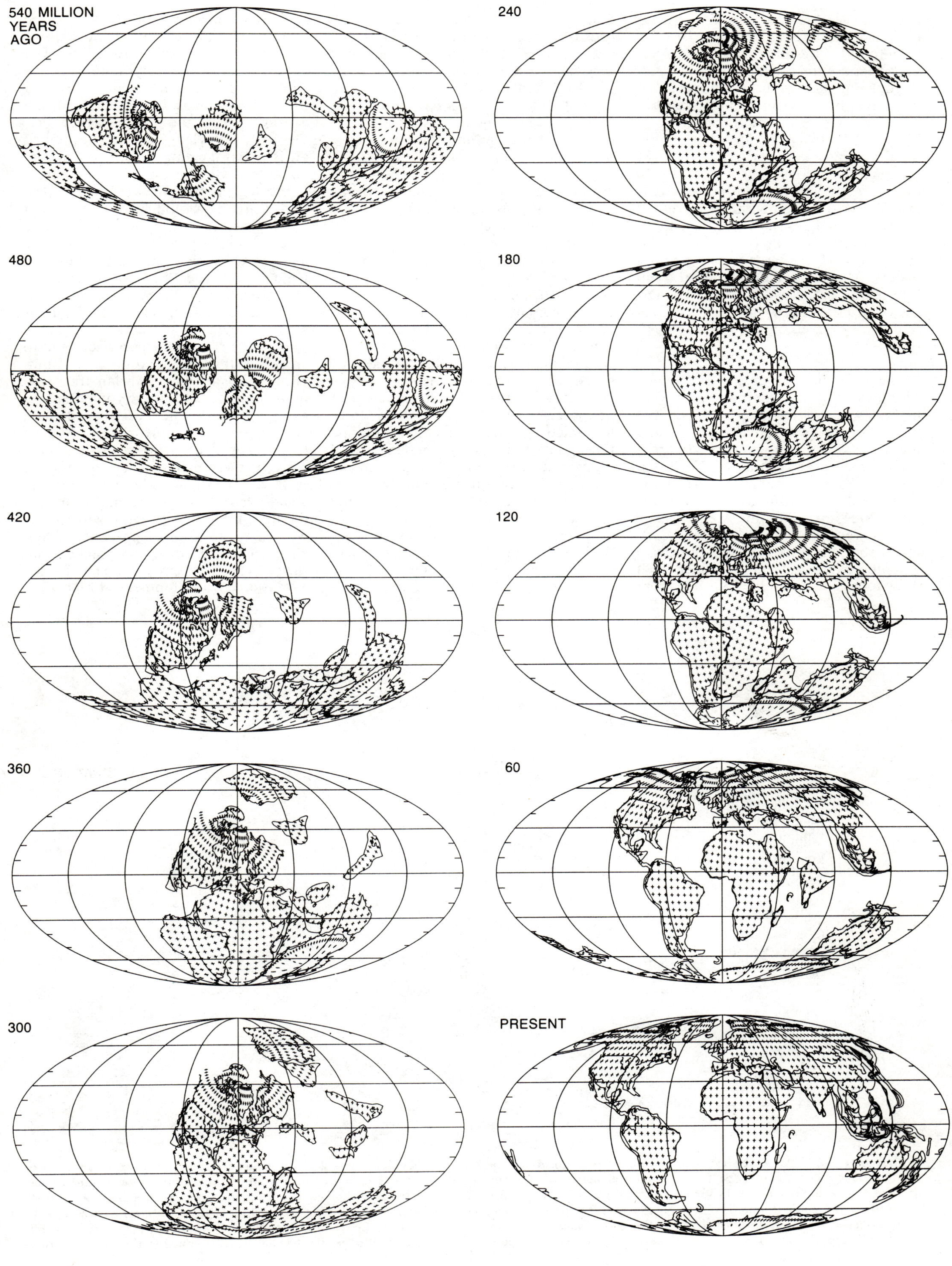
540 MILLION YEARS AGO
240
480
180
420
120
360
60
300
PRESENT

tween these early rocks and more recent ones. Although no fossils have been found in the Greenland rocks (perhaps because the rocks have been so altered by metamorphism), evidence of primitive life is found in somewhat younger ones [see "The Biosphere," by Preston Cloud, page 95]. Some of the basalts found in the earliest earth rocks have compositions that reflect much higher melting temperatures, as if the rate at which the temperature increased with depth in the earth were much greater then than it is now; this is not surprising in view of the earth's early thermal history. Moreover, before about 2.5 billion years ago there were few large masses of granitic rocks and the kinds of sediments formed on shallow continental shelves.

The differences indicate a machine whose internal temperatures were higher, whose atmosphere and oceans were deficient in oxygen and whose land masses were isolated, smaller areas rather than large continents. Yet at all times the average compositions of the rocks and their similarities from one place and time to another demonstrate that the recycling machine was operating at something like the same rate and with the same general character as it does today. It undoubtedly underwent gradual changes, and there are also records of abrupt episodic shifts. The first glaciation of the earth was recorded early in Precambrian time. A more significant change in the cycle came about 2.5 billion years ago, when there seems to have been a sudden rise in the production of granite and the appearance of large continental shelves.

Later in the Precambrian, about a billion years ago, the machine began to look much as it does today. Oxygen in the atmosphere started increasing as photosynthetic organisms grew in numbers and efficiency, and as the march toward more complex life forms accelerated. Although the land surfaces were populated only by algal, fungal and bacterial species, rock weathering in Precambrian soils and the formation of river and lake deposits were proceeding much as they are today; their rates, however, may have been lower. The interior of the earth had by now settled down to something like its present state and through partial melting some regions of the mantle were being differentiated and depleted in certain elements with respect to the mantle as a whole. Whether plate tectonics was the primary mode of heat loss at that time is an open question. The magnetism of ancient rocks gives evidence of continental drift and reversals in the polarity of the earth's magnetic field, and so the dynamo in the earth's core that generates the field must have been in steady operation [see "The Earth's Core," by Raymond Jeanloz, page 13]. Perhaps the plates were too thin to act quite as they do today, or the average size of the plates was smaller. Geologic knowledge of the Precambrian is too fragmentary to give firm answers to these questions yet.

Most geologists working in the past two centuries would have agreed that the biggest change of all was the one at the boundary of Precambrian time and the Cambrian period, marking the beginning of the Phanerozoic eon: the "known" part of the geologic record. It was then that organisms with shells evolved; their fossils make it possible to date the rocks more precisely and to erect the stratigraphic time scale. Compared with the Precambrian there are many more areas of these younger rocks exposed in an unmetamorphosed state, and so one can more readily deduce the course of earth history since then. This change, which is generally believed to have taken place about 570 million years ago, was far more important for the life of the planet than it was for the working of most other parts of the machine, which by then had reached maturity. The parts most affected by the evolutionary development of higher or-

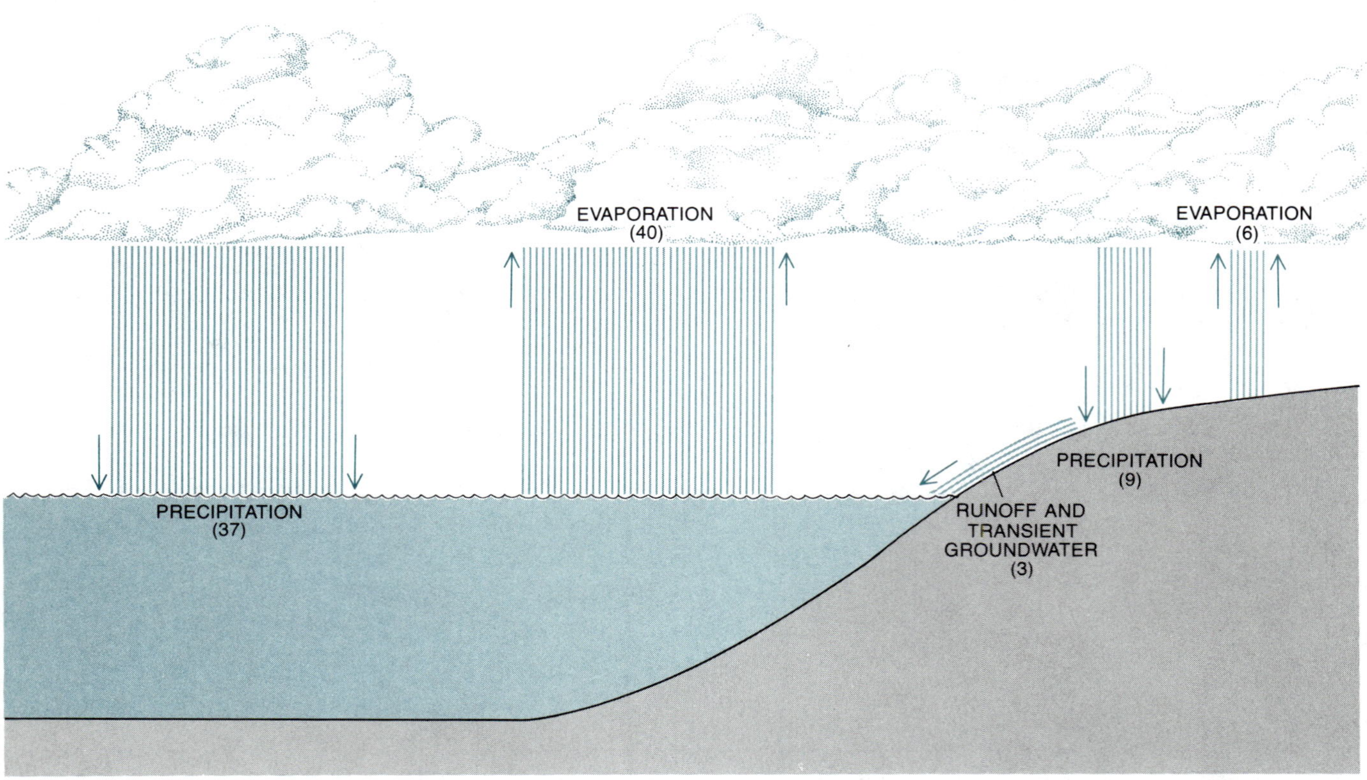

HYDROLOGIC CYCLE, represented quantitatively in this diagram, serves as a model of the cyclic flow of material between different parts of the earth. The numbers shown are multiples of a basic unit equal to 10,000 cubic kilometers of water per year. In spite of occasional short- and long-term perturbations, the global balance among the world's major water reservoirs—atmosphere, oceans, continents and polar ice caps—is quite steady from year to year. In addition to the amounts indicated here water is also brought to the surface of the earth by volcanism. This comparatively small increment consists of former surface water recycled through the interior by way of subduction zones as well as "juvenile" water, namely part of the original water of the mantle that has never escaped to the surface before.

ganisms were those at the surface, where the chemical influence of life processes is felt. The atmosphere arrived at something like its present oxygen level; the deposits of limestone became dominated by the shells of shallow marine organisms, and the chemical cycles of carbon dioxide, oxygen, phosphorus and nitrogen rather abruptly shifted to their present states.

Since that time changes in the rates and quantities of sediments controlled by organisms have been determined by the course of evolution. The vascular land plants first appeared about 400 million years ago and coal beds began to be widespread. The evolution of the flowering plants came about 120 million years ago. The siliceous algae (diatoms) and the pelagic foraminifera evolved about 130 million years ago and started to form the abundant siliceous and calcium carbonate oozes of the sea floor. Much biological sedimentation was shifted to the open sea floor, whose evolution can be understood only for the past 150 million years. The older record of sea-floor spreading has been swept into subduction zones and destroyed by metamorphism or melting.

The earth in old age is predictable from the present running of the machine. As the original supply of radioactive elements steadily decreases, the internal heat of the earth will diminish and simple heat conduction will slowly take over from the currently dominant mode of heat transfer by convection. As the earth cools, the rigid plates of the outer crust will thicken, and eventually they will congeal and become motionless. The "hot spots" that form volcanic centers will cool and solidify. With no internal forces to throw up mountains and move continents around, the external machine on the surface will take over, reducing much of the surface elevations of the land to plains elevated slightly above sea level. Sediment, the debris of erosion, will spread over land surfaces and sea bottom with no interruption by deep-seated rocks brought to the surface. A new balance of chemical elements based on a static tectonic system will lead to a shift in the composition of the atmosphere and the ocean, finally leading, as all mountain building and sedimentation ends, to a new steady state determined almost entirely by life's balance between photosynthesis and respiration.

That balance itself, together with the total mass of biological material, might change significantly as the nutrient reservoirs of the ocean and the atmosphere interact only with the thin surface skin of the earth. This brief essay at describing the future workings of the earth machine is enough to show how dependent the present workings of the planet are on interactions with the interior. No one

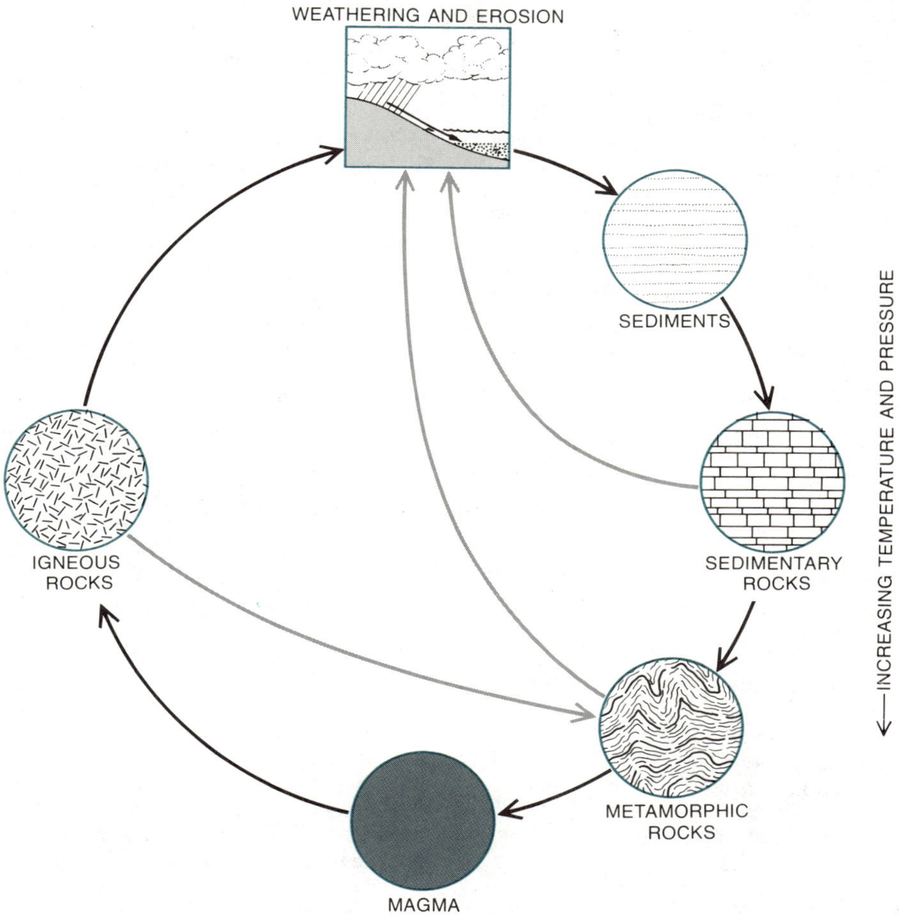

ROCK CYCLE, first proposed by Hutton almost 200 years ago, is still the basis of the geologist's view of the changing earth. In Hutton's version rocks are weathered to form sediment, which is then buried. After deep burial the rocks undergo metamorphosis and/or melting. Later they are deformed and uplifted into mountain chains, only to be weathered again and recycled. The modern theory of plate tectonics is in a sense an elaboration of Hutton's rock cycle.

is sure, however, of how such perturbations in the operation of the present machine can be predicted quantitatively, as is suggested by the current experience with carbon dioxide.

It has now been almost 100 years since Svante A. Arrhenius, the great Swedish chemist, called the attention of earth scientists to the effects of carbon dioxide on climate and its possible relation to glaciation. For much of that period some workers have been concerned about the steadily rising amounts of carbon dioxide in the atmosphere contributed by the burning of fossil fuels. As the pace of carbon dioxide emissions has increased dramatically in the latter half of the 20th century and as knowledge of the carbon cycle has grown, the issue has come to the point where government commissions and other national bodies estimate how much climatic change there is likely to be and what its effects might be. The comparatively small changes man has worked in the earth machine might have enormous consequences. Large as the earth is, the behavior of one of its biological species may be enough to seriously disturb its balance.

In order to estimate the effects of increased carbon dioxide in the atmosphere one must consider all the fluxes and reservoirs. The human disturbance has been to change a single flux by increasing through coal mining and oil production the rate of return of buried organic carbon to the surface by orders of magnitude over the estimated preindustrial flux. To forecast the consequences one must follow through the entire system the changes that result. More carbon dioxide in the atmosphere leads to a slight increase in global temperatures as a result of the "greenhouse effect." An additional dissolution of calcium carbonate in the ocean and imperceptible changes in the balance of weathering and sedimentation of silicate and carbonate minerals will follow. The warming of the atmosphere will lead to some melting of the polar and glacial ice and a rise in the sea level, and very probably a shift in climatic belts. Reciting this list is only the beginning of a consideration of the innumerable ramifications of a shift in only one flux in the system. It is this complexity that makes the estimation of the ultimate effects so difficult.

Burning fossil fuels is only the most recent perturbation of the earth machine. Geologic history is in one sense a recital of the many small shifts of the balance, some local and some global, that have characterized this otherwise smooth-running mechanism. The closer one looks at any complex machine, the more clearly one can detect fluctuations in its operation. In the midst of thousands of fluctuations that generate background noise—small shifts in sea level, plate-spreading rates and erosion rates, among others—one can see the signal of large and unusual events, infrequent and sometimes catastrophic. Near the end of the Miocene epoch, about 11 million years ago, the convergence of the African and European plates closed the mouth of the Mediterranean Sea. The Mediterranean dried up and thick layers of evaporites, principally salt and gypsum, were laid down. Soon afterward continuing plate motions reopened the Mediterranean to the Atlantic and it refilled. An immense waterfall existed for a time as water from the Atlantic spilled in to fill up the entire Mediterranean basin. Spectacular as the event was for this area of the world, it probably had a negligible effect on the global cycle; the evaporated water and salt removed from the ocean was minor compared with the huge volume of water in the world ocean.

Salt deposition at another time in the history of the earth was a part of a much more pervasive global cycle, one familiar to all geologists for its relation to massive extinctions of many biological species and almost complete withdrawal of the seas from every part of the continents. The end of the Paleozoic era, the boundary between the Permian and the Triassic periods, about 225 million years ago was marked by the complete assembly of the supercontinent Pangaea with the superocean Panthalassa surrounding it. Over a period of about 200 million years plate motions had gradually joined various pieces of continents into a single land mass. In the process evaporation from the narrow gulfs and bays of closing oceans and from arid regions led to great amounts of salt being withdrawn from the oceans, decreasing the salt content of the open ocean slightly and perhaps altering its density-driven circulation.

As the continents joined, their total perimeter shrank. As a result of mountain building caused by continental collisions in the course of the assembly the continents were dominated by high mountains. The ocean basins were enlarged by the slowing of plate-spreading rates and the lowering of midocean ridges as plate motions became largely limited to the oceanic crust. The consequence of all these effects was a severe restriction of the area of the continental shelves. The diminished habitats of the shallow marine populations of the world and more variability in the climatic regimes of the land, which had recently gone through another glacial age, set the stage for the rapid evolution of new species and the extinction of old ones. A lot of guesswork is needed here because the rock record for the period is sparse. Most of the sediments deposited on the continents have since been eroded and the sea-floor record has been swept into the subduction zones.

Even this unusual time was soon superseded by a return to normal conditions as the supercontinent proved unstable. It began to rift into separate continental masses, and the Atlantic and Indian oceans opened. The seas transgressed the continents, salt beds were eroded, the salt returned to the ocean and the normal pattern of life resumed on the continental shelves. The episode demonstrates the machine's ability to right itself and settle down to a smooth operation after a violent shudder.

The glacial ages represent another kind of perturbation, one that primarily affected the surface and near-surface layers of the earth. The earth has been glaciated several times in its history, at least twice during the Precambrian, once in the early Paleozoic, once near the end of the Paleozoic and many times in the Pleistocene, which ended only about 10,000 years ago. Each of these events may have been triggered by continental-drift patterns that put a continent in a polar region and caused shifts in oceanic circulation. Perhaps there was a correlation with a wider than normal fluctuation of the carbon dioxide level. In each glacial period the periodicity of changes in the earth's orbit around the sun led to oscillations in the advance and retreat of the ice; the oscillations, called Milankovitch cycles, are now thought to account for the glacial and interglacial stages of the most recent ice ages. Each time, the postglacial earth returned to its preglacial state with little change, leaving behind a record of glacial sediment and grooved and striated bedrock, the pavement over which the ice moved.

In the past few years geologists have begun to evaluate disturbances of the earth's balance due to extraterrestrial sources. A thin bed of iridium-enriched clay has been found in a number of places around the world where sediments were deposited during the transition from the Cretaceous period to the Tertiary, about 65 million years ago. It has been proposed by Luis W. Alvarez, Walter S. Alvarez, Frank Asaro and Helen V. Michel of the University of California at Berkeley that this bed is the product of the impact of a large meteoritic body. Among the postulated consequences of the collision (with a body estimated to weigh about 10^{18} grams) were clouds of dust that obscured the sun, sudden cooling of the atmosphere and the ocean, changes in the chemistry of the atmosphere and the deposition of a thin layer of clay enriched in certain rare metals such as iridium. These changes are thought to have led to the extinction of many plant and animal species in the sea and the dinosaurs on the land.

Regardless of the durability of this hypothesis as an explanation of the geochemical and biological associations of the boundary between the Cretaceous and the Tertiary, geologists have come to recognize that collisions with asteroid-size bodies have been an important feature of the earth's history since its formation 4.6 billion years ago. Such heavy impacts may have come every 100 million years or so. Most of the events have left little trace because of erosion over the next few tens of millions of years, but some have left conspicuous circular scars.

As the interconnectedness of the huge and complex earth system becomes more widely recognized, earth scientists are learning how to link up subsystems and explore new relations. Oceanographers have combined marine chemistry, marine biology and the geology of the sea floor, from midocean ridges to volcanic-island arcs, to see the oceans as an entire system. Indeed, it is the only way to estimate the interactions of the oceans with the carbon cycle and the carbon dioxide greenhouse [see "The Ocean," by Wallace S. Broecker, page 69]. With geologists and paleontologists they are now mapping the past of the oceans, particularly in relation to the recent glacial epoch. Predicting the operation of the surface part of the machine, powered by solar radiation, has become the business of oceanographers, geologists, atmospheric scientists and biologists. The ability to predict is dependent on devising ever more inclusive and sensitive models of the system and testing them against observations. Geophysicists working on the dynamics of the interior heat engine are engaged in correlating its structure and constitution, as revealed by earthquake waves, with its mineralogical and chemical constitution, as deduced from laboratory experiments on rock systems at high temperatures and pressures. Isotope-ratio studies of rocks derived from the interior and brought to the surface provide observations that tell about mixing mechanisms in the mantle. Theoretical studies of convection dynamics in plastically deformable solids at high pressures are showing how the mixing proceeds. The core, the mantle and the crust are no longer regarded as separate domains but as interacting parts of a larger system whose properties and dynamics are modeled by geophysicists, geochemists and petrologists.

Earth scientists have long been familiar with the ways in which the actions of the interior affect the exterior through volcanism, mountain building, the flow of heat and the geomagnetic field. Now they are seeing how chemical weathering and the differentiation of the materials of the interior as they are brought to the surface react on the interior as the altered material is plowed back into the mantle by subduction. In this way the surface machine is further coupled to the interior machine. As the subsystems are linked the earth may come to be thought of more in the ways one thinks of a highly differentiated organism: as a system so complex that the ultimate reduction to simple forces and bulk compositions does not lead to a satisfactory understanding of the wonderful diversity and detail that can be observed directly at the surface and sensed remotely in the interior.

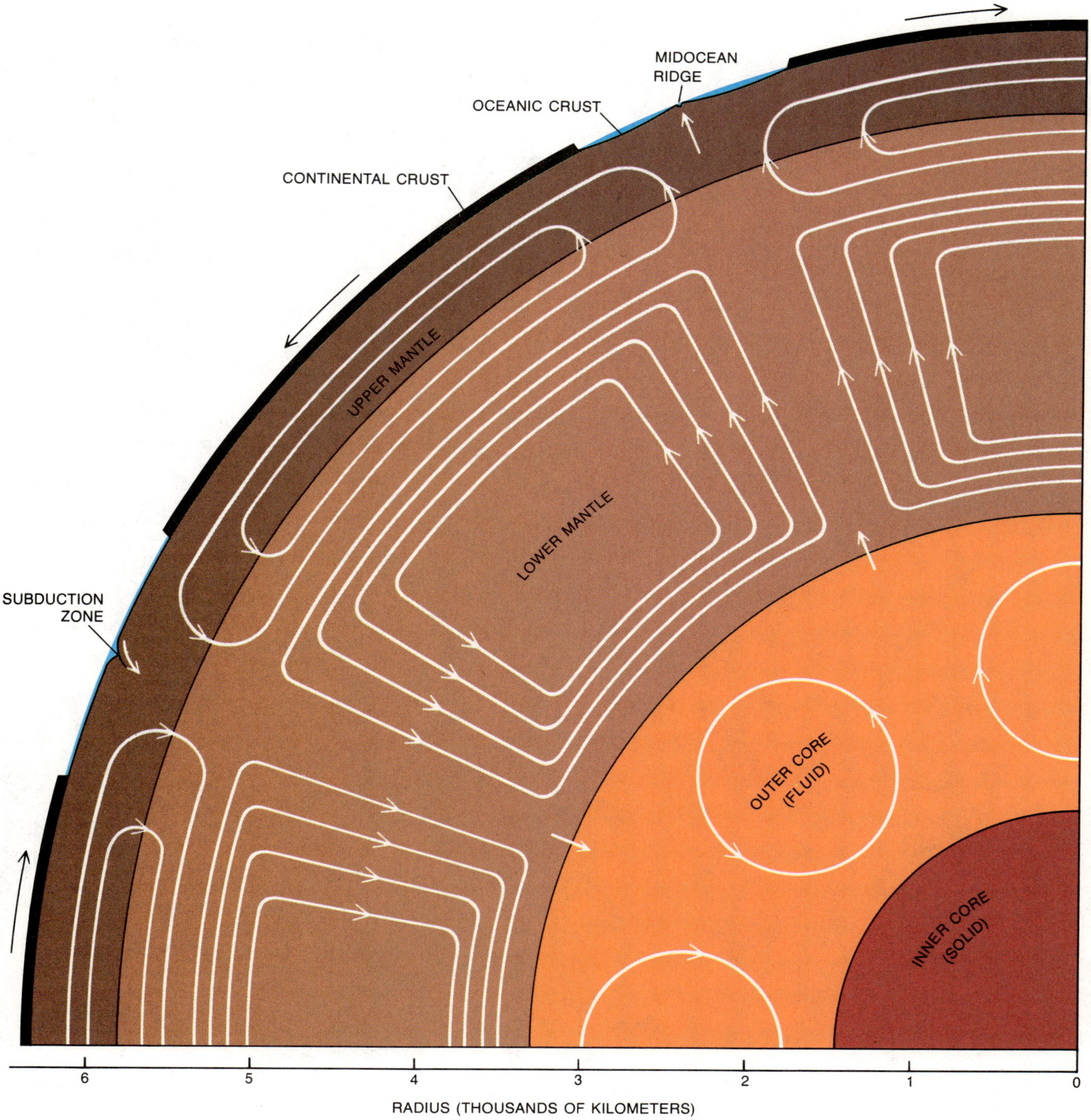

LARGE-SCALE MOTIONS of the major parts of the earth are indicated by arrows in this highly schematic diagram. Heat-driven convection in the fluid outer core has a dynamo effect that is responsible for the geomagnetic field. Convection in the upper mantle drives plate tectonics. Volcanism transports molten material to the surface at mid-ocean ridges and other places. Solid material is returned to the interior at subduction zones. The degree of mixing between the upper mantle and the lower mantle is a subject of debate; in this case a model calling for separate convection cells has been adopted. Mixing of material between the lower mantle and the outer core is still speculative.

2

THE EARTH'S CORE

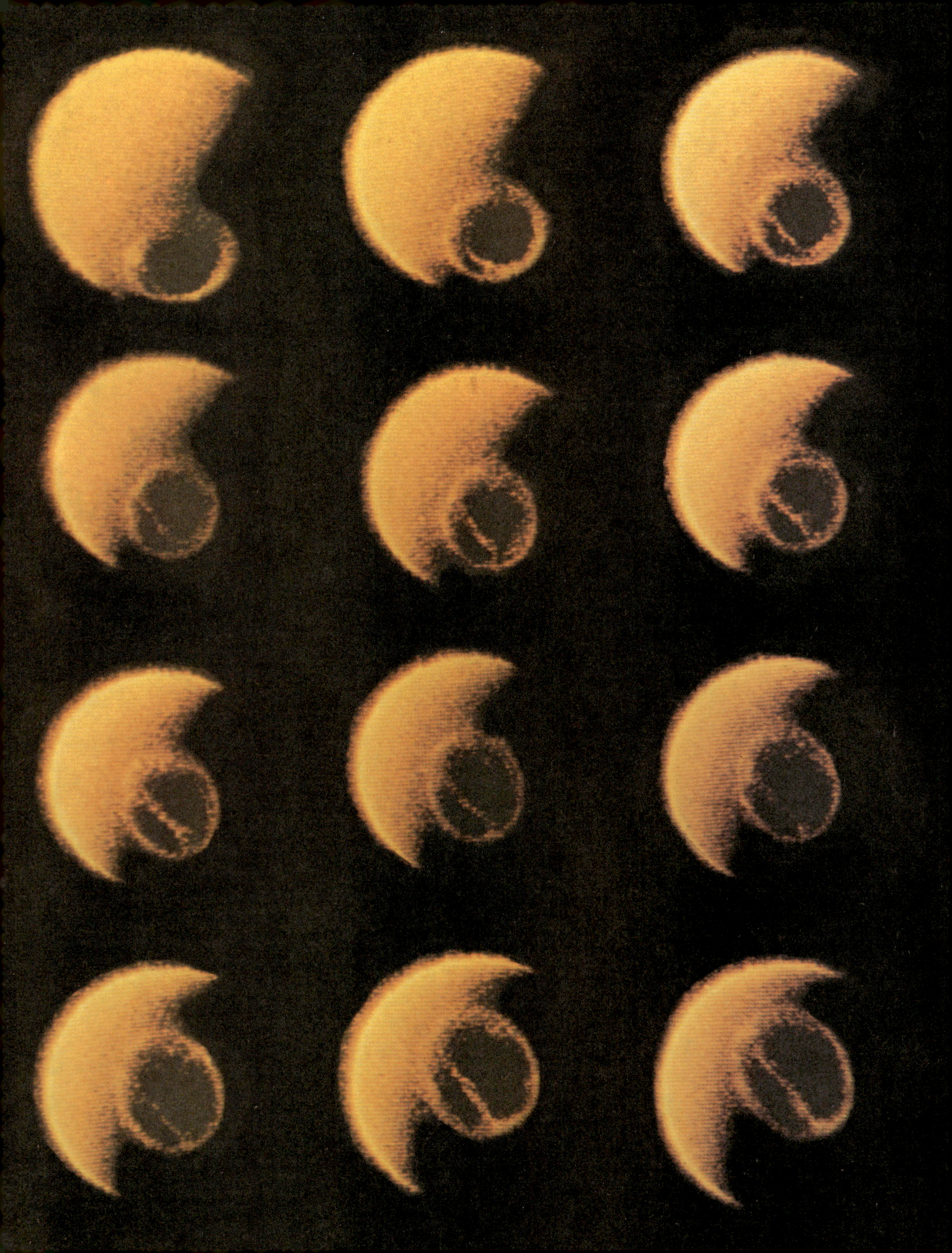

The Earth's Core

by RAYMOND JEANLOZ

Indirect evidence indicates that it is an iron alloy, solid toward the center but otherwise liquid. It is the turbulent flow of the liquid that generates the earth's magnetic field

It is ironic that one of the great spectacles of terrestrial nature, the light of the aurora shimmering in the night sky, is a clue to the character of the earth's hidden and enigmatic core. The ultimate cause of the aurora is the interaction of the magnetic field generated in the core and the "wind" of electrically charged particles flowing outward from the sun. The core also has much to tell about the earth's formation and geologic history. Indeed, there are indications that it may still be influencing the distribution of temperature in the overlying mantle and thus may indirectly govern large-scale geologic processes at the surface. It is also clear that the composition of the core is a major factor in any model of the bulk chemistry of the earth.

The present nature of the core is best determined from seismological data, that is, the information gathered by studying the acoustic waves generated by earthquakes. Such data reveal that the core extends from a depth of about 2,900 kilometers (1,800 miles) to the center of the earth at 6,370 kilometers (3,960 miles). They also show that the inner core is solid, with a radius of about 1,200 kilometers, and that the outer core is liquid. What the composition of the solid and the liquid might be are matters to which I shall return.

Conditions are known to be extreme at these depths. The pressure ranges from 1.3 to 3.5 million atmospheres, which is to say from 1.3 to 3.5 million times the atmospheric pressure at the surface of the earth. Temperatures are estimated to be in the range from 4,000 to 5,000 degrees Celsius (7,200 to 9,000 degrees Fahrenheit).

The most direct information on the core in the past comes from studies of the ancient magnetic field of the earth. For example, the magnetization of some of the oldest rocks suggests that whatever process generates the magnetic field in the core was already operating 3.5 billion years ago. Students of such matters are still far from a thorough understanding of the geomagnetic field: how it is generated, when and how it started and how it has evolved. Nevertheless, it is now possible to begin to understand how the core might have formed and evolved and how these processes may have affected the geologic evolution of the earth.

The geomagnetic field is often described as being like that of a dipole. In other words, it looks like the field that would be generated by a bar magnet at the center of the earth, with the lines of force looping from the south magnetic pole to the north. This is actually a parochial description, because the magnetic field is like that of a dipole only near the surface of the earth, where it can be studied most readily.

Around the earth in the magnetosphere the lines of force in the magnetic field are strongly distorted by the solar wind; they are crushed toward the earth on the side facing the sun and swept far into space on the night side of the planet. Similarly, all models of the source of the magnetic field show distorted field lines in the core. Nevertheless, it is worth remembering that about 90 percent of the field now observed at the surface is dipolar. The rest consists of a more complex pattern of field lines that can be described in terms of several poles rather than just the two that describe most of the present field.

Since the pioneering work of Walter M. Elsasser of Johns Hopkins University, Edward C. Bullard of the University of Cambridge and others the geomagnetic field has been understood as originating with magnetohydrodynamic processes in the earth's liquid outer core. In general terms the processes entail convection in an electrically conducting fluid, with the result that the core acts as a dynamo maintaining and regenerating the magnetic field. Specifically, as the field lines directed toward the center of the earth (the poloidal lines) enter the outer core they are pulled in the direction of the earth's rotation. The rotation of the solid inner core probably tends to wrap the field lines around the earth's axis (producing a toroidal component).

Furthermore, it is thought the field lines become contorted by smaller-scale cyclonic motions that result from the fact that the core is rotating along with the rest of the earth. The cyclonic motions are analogous to the hurricane patterns that arise in the atmosphere. Although the contortions of the magnetic field lines play a central role in current theories of the magnetic dynamo, neither the exact origin nor the detailed pattern of the contortions is known. It is worth emphasizing, however, that without the dynamo process the magnetic field would certainly die out within 10,000 years or so. Therefore the field must be continuously maintained or regenerated by the fluid motions.

In order to better understand the nature of the magnetic field within the core it is necessary to know the convective-flow pattern of the liquid. The trouble is that the magnetic field can significantly modify the flow generating the field in

MAGNETISM OF THE EARTH is visible in these images of the aurora made from data transmitted by the spacecraft *Dynamics Explorer 1*. The aurora is the circle at the lower right in each image. It is created when electrically charged particles in the "wind" of the sun's expanding atmosphere are trapped in the earth's magnetic field. Plunging into the earth's atmosphere along the magnetic lines of force descending toward the magnetic poles, the solar-wind particles interact with molecules in the earth's atmosphere to generate the light of the aurora (here actually detected at ultraviolet wavelengths). The bright area at the upper left in each image is the side of the earth lighted by the sun. Each image represents data received for a 12-minute period. The images show a hitherto unknown configuration of the aurora, in which the circular area of emission is crossed by a linear one; the configuration is called the theta aurora because of its resemblance to the Greek letter. The images were obtained with the University of Iowa's auroral-imaging instrumentation. They are provided through the courtesy of Louis A. Frank.

the first place. Hence no one has yet fully solved the problem of determining the fluid motions in the outer core.

A logical starting point, however, is to first ask what the flow would look like if there were no magnetic field. Possible answers to this purely hydrodynamic question are emerging from the theoretical and experimental work of Friedrich H. Busse and his colleagues at the University of California at Los Angeles. They find that both the presence of the solid inner core and the rotation of the earth are major influences on the flow pattern in the liquid outer core. In conditions thought to be appropriate to the earth's core the pattern takes the form of screwlike rollers. One can speculate that if this pattern of flow is not too strongly modified by the presence of the magnetic field, the rollers are intimately associated with the contortion of the field lines that is required for the dynamo.

The magnetization that is retained in rocks of different ages has been studied by, among others, M. W. McElhinny and his colleagues at the Australian National University. They find that a geomagnetic field of about the same intensity as the present one existed at least 2.5 billion years ago and probably existed 3.5 billion years ago. (The latter determination was made on the basis of only a few samples.) Although data are absent for a substantial fraction of the earth's history, inasmuch as the planet is 4.6 billion years old, it seems that the geodynamo started operating fairly early in geologic time.

The fact that the presence of an inner core appears to be important in the process generating the magnetic field leads to the further inference that this solid region at the center of the earth has existed for at least 3.5 billion years Even though the inner core may have grown or shrunk since then, there is no reason to infer that the general nature of the flow in the outer core has changed substantially throughout recorded geologic time. In other words, the magnetic evidence suggests that the basic structure of the liquid outer core surrounding the solid inner core has existed for at least three-fourths of the earth's history.

The understanding of the geomagnetic dynamo and of the convective-flow pattern in the core has advanced considerably in the past several years. It is reasonable to predict that the theoretical and experimental work now being done will lead to major new insights. Nevertheless, certain aspects of the problem are still baffling.

A good example is the reversal of the magnetic field that has occurred hundreds if not thousands of times in geologic history. In a reversal the north magnetic pole switches from pointing toward geographic north to pointing south (or vice versa). Reversals appear not only in the earth but also in the sun and even in laboratory dynamos, presumably in response to the chaotic nature of magnetohydrodynamic processes. On a practical level the frequent and apparently random geomagnetic reversals have provided an exceedingly useful clock for timing geologic events and correlating geologic deposits. In-

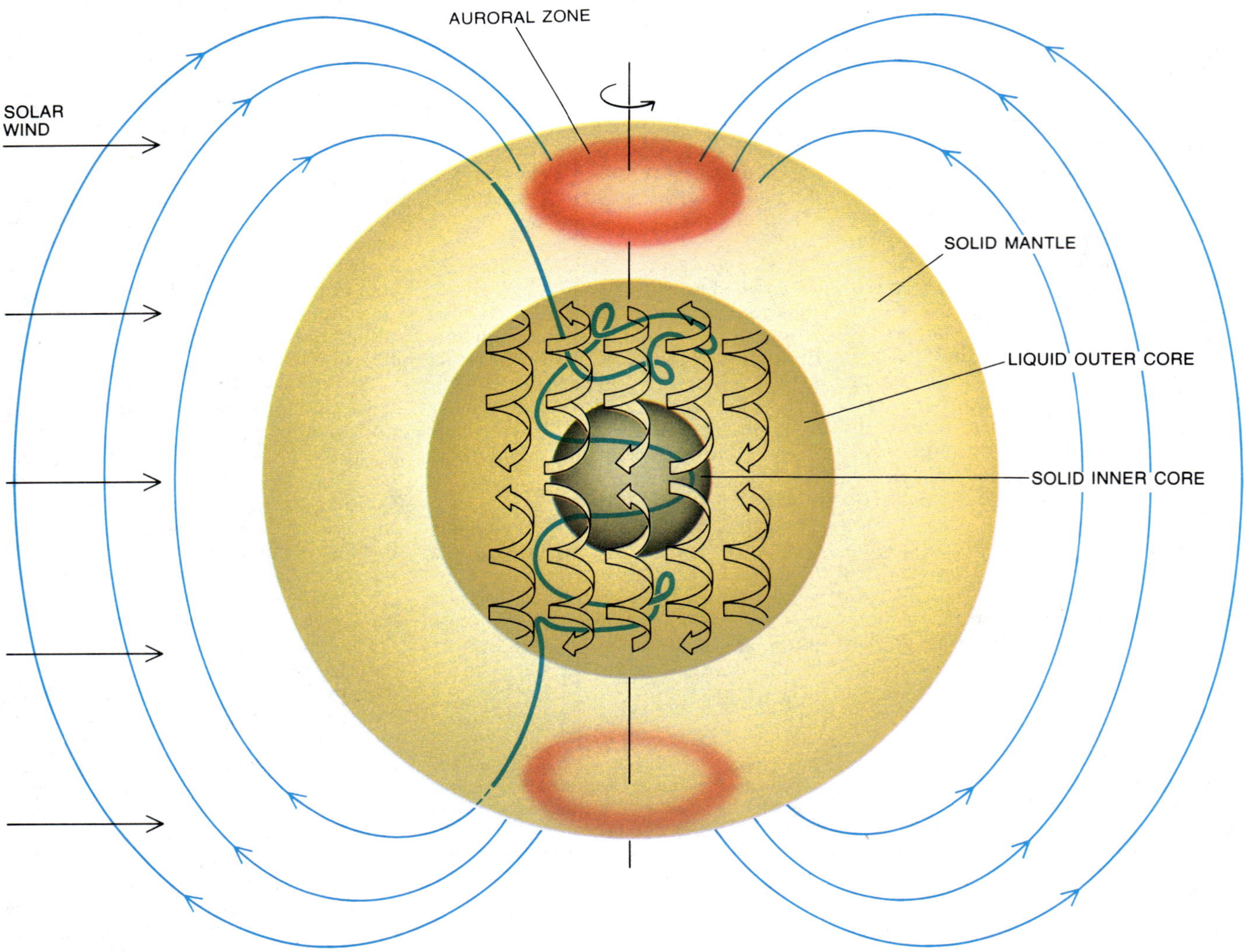

MAGNETIC FIELD IS GENERATED by a dynamo in the core. The details of how the dynamo works are not known, but in the model depicted here it is assumed that the electrically conducting metallic liquid of the core flows in screwlike rollers. The lines of force in the magnetic field would be threaded through the rollers; here a single such line is depicted. It is the thick line from north to south.

deed, it is the magnetic reversals recorded in the rocks of the ocean floor that have yielded one of the main lines of evidence for the theory of plate tectonics.

It is only fairly recently that high-quality data have become available to trace in detail what happens when the magnetic field reverses. Apparently the magnetic pole follows a convoluted path at the earth's surface. It is clear, however, that the field can reverse completely within just a few thousand years.

Several models have been proposed for what the field should look like during a reversal. According to one of them, as an example, either the north or the south magnetic pole should appear to follow a line of constant longitude, with the pole crossing through any given sampling location on the surface of the earth in the course of a reversal. One can visualize the magnetic pole as splitting into a hoop that sweeps the surface in a north-south direction. The data now available, however, do not support this hypothesis.

A major difficulty in modeling the field as it reverses is that during the process the absolute intensity of the field is reduced to about 10 percent of its normal value [*see illustration on next page*]. At present the geomagnetic field is about 90 percent dipolar, the rest being multipolar. No one is certain what fraction of the reduced field is dipolar while a reversal is under way. To derive the poles shown in the illustration it is necessary to assume that the field was dipolar during the depicted reversal. Perhaps, however, there are effectively several magnetic poles existing simultaneously during a reversal. If they fluctuate in strength, they can give rise to the erratic path seen in the illustration. This possibility can be investigated only by studying the same reversal at several different places.

Reversals seem to be more than just a passive dying out and rebirth of the field because they appear to happen on a time scale that is much shorter than the time it would take for the field to be regenerated. Hence one may ask: Are reversals caused by turbulence or shifts in the detailed flow patterns in the outer core? No one has been able to answer the question. It is quite possible the dynamo is self-reversing, that is, a reversal can be initiated internally, without any external trigger.

Regardless of the mechanism, one consequence of reversals is only now beginning to be recognized. What is being found in the growing new discipline of biomagnetism is that complex organisms synthesize magnetic components whereby their behavior can be significantly affected by changes in the geomagnetic field. Therefore one may speculate that reversals play a role in biological evolution.

The geologic record of the earth's magnetic field yields information on the nature of the core in the past. In order to proceed further in understanding the core it is necessary to consider the available data bearing on its present nature as well. The only detailed and direct information comes from seismological studies, which provide values of the density of the material and the velocity and attenuation of sound as a function of depth in the earth. "Sound" is employed here in a loose sense to refer to the mechanical waves produced by earthquakes (or large manmade explosions) and propagated through and around the globe. Such waves are of low frequency compared with audible sound: from about 10^{-4} to 10 hertz, or cycles per second, which is some 100 to a million times lower in "pitch" than the concert A. From these data the pressure can be calculated at each depth. Moreover, liquid and solid regions can be distinguished because a liquid transmits only compressional waves (waves moving back and forth in the direction of their travel), whereas a solid transmits both compressional waves and shear waves (waves moving at right angles to their direction of travel).

The basic structure of the earth implies a crude but strong constraint on the deep temperature. Evidently the geotherm (the average temperature as a function of depth) is below the melting point of the inner core and the mantle, since they are both solid. By the same token the geotherm is above the melting point of the outer core.

Usually this argument is taken one step further on the assumption that the inner core is solidified outer-core material. The assumption is plausible and is not contradicted by any data now available. No one has yet demonstrated, however, that the inner and outer core are in chemical equilibrium, as the assumption would imply. In fact, the inner core may be in chemical disequilibrium with the outer core, which in turn may be (and is often taken to be) in disequilibrium with the mantle. In other words, the composition of the inner core may not be related in any direct way to the composition of the outer core.

If the inner core has formed (or is still forming) by crystallization from the liquid outer core, the boundary between them is fixed within the temperature interval of melting and solidification of core material at a pressure of 3.25 million atmospheres for the present size of the inner core. For example, the inner core could be pure iron that has crystallized out of the alloy liquid of the outer core. The temperature at the boundary between the inner and the outer core is similarly constrained even if the outer core is heating up and hence is growing by the melting of the solid inner core. In either case one would in general expect a region of partial melting between the inner and the outer core, because only the simplest chemical systems melt at a sharply defined temperature, that is, begin melting or begin solidifying at the same temperature.

The core is thought to be a complex alloy, which would therefore melt over a range of temperatures. As a result there has been much interest in recent seismological studies by V. F. Cormier of the Massachusetts Institute of Technology and G. L. Choy of the U.S. Geological Survey, who find evidence for a "mushy" zone of reduced velocity and relatively high attenuation of seismic waves in the top few hundred kilometers of the inner core. Their conclusions are uncertain because of the great difficulties involved in resolving such details of physical properties near the center of the earth. It does appear, however, that there may be an anomalous zone such as one would expect for a slurry of liquid and crystals. This is circumstantial evidence that the top of the inner core is right at the melting point. Thus if the composition of the core were known, an experimental determination of the melting point of this material at the pressures of the core would yield a direct determination of the temperature near the earth's center.

It is also possible to infer that the temperature has not varied enough to completely melt (or solidify) the core within the past 2.5 to 3.5 billion years. The inference is based on the record of the earth's magnetic field suggesting the inner and the outer core have existed for at least that length of time.

The core is usually represented as consisting mainly of iron. This interpretation is clearly in agreement with the seismological data, but two other lines of evidence can be invoked to strengthen the conclusion. The first is that the generation of the magnetic field requires the core to be metallic (that is, electrically conducting) in order for the geodynamo to operate. The second is that no other element having the observed properties of the core is abundant enough in the cosmos to be a plausible candidate.

Hence there is a gross separation of the earth into an iron-rich region (the core) and a silicate region (the mantle and the crust). Silicates are the complex silicon-oxygen compounds making up rocks, so that the two regions are reminiscent of the two general classes of meteorites: iron and stony. Beyond this, however, there is no evidence for a more detailed analogy between the earth's core and the nature of meteorites. In fact, it is not possible that the core is made up only of iron or the nickel-iron alloy commonly observed in iron meteorites. This is evident from a comparison of the density of iron alloys at high pressures with the density of the core: a small amount of a component less dense than iron, such as sulfur, oxygen or silicon, must also be present.

There is no consensus on the composition of the core other than that it is predominantly iron. In part the reason is that gross physical properties, such as the measured densities and seismic velocities within the core, cannot be exploited to uniquely determine the core's chemical composition. Moreover, only a small amount of any of the alloying elements that have been proposed is required in order to match the observed properties of the core. (Typical values are about 8 to 10 percent by weight.) Finally, it is quite possible—probable according to many investigators—that the core alloy has many components in addition to iron.

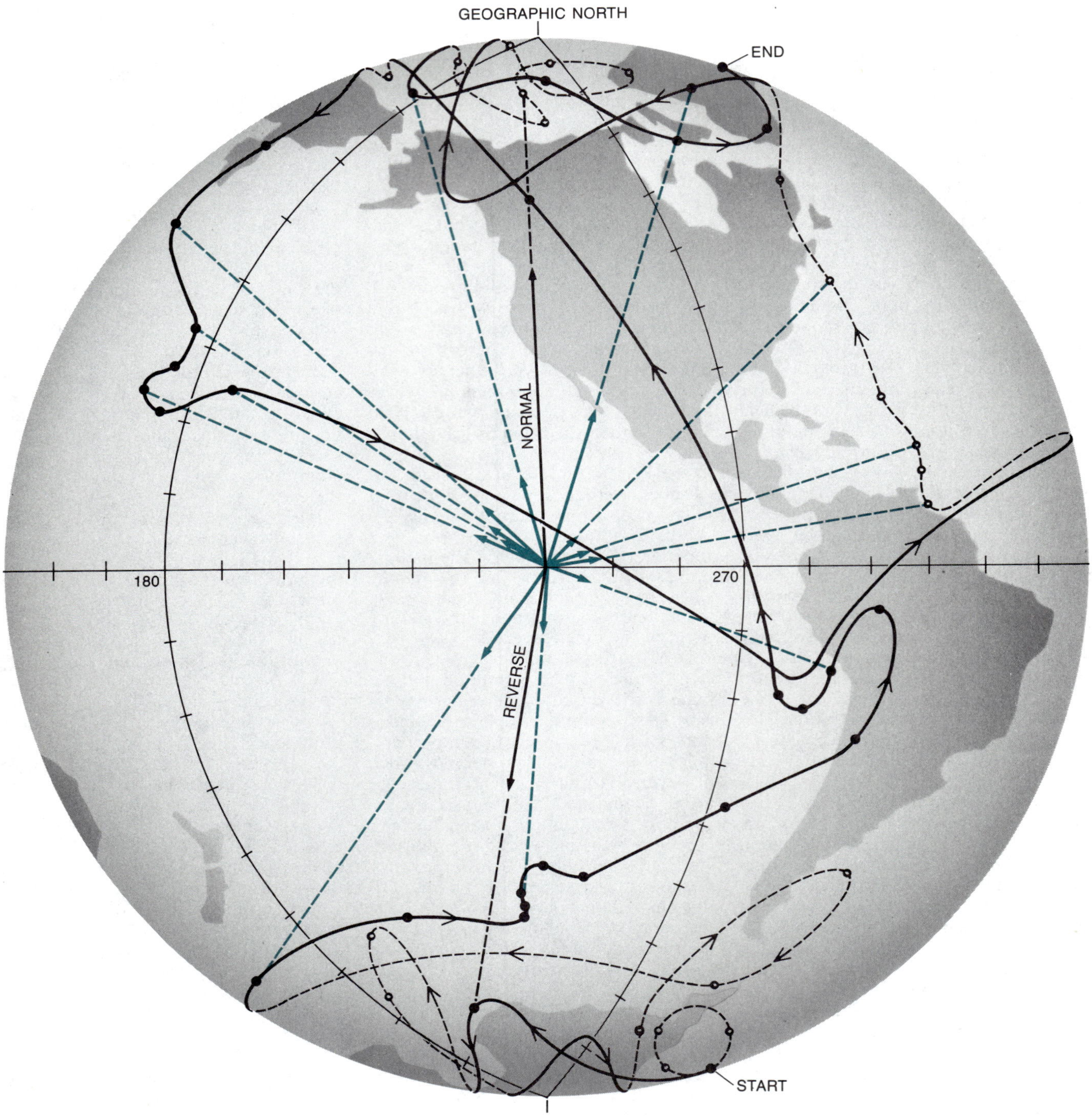

REVERSAL OF THE GEOMAGNETIC FIELD gives clues to the action of the dynamo in the core. A reversal that took place 15 million years ago is traced here on the basis of the magnetism recorded in a sequence of lava flows at Steens Mountain in Oregon. The switch was from reverse to normal, with normal meaning that magnetic and geographic north are in the same direction. The north magnetic pole followed a convoluted path, recorded here for a period of 15,000 years; each filled circle represents a separate measurement of the pole position. The field's strength and direction are indicated at approximately 500-year intervals by the colored arrows. During the reversal the geomagnetic field did not necessarily have two poles, as it does today, so that the poles depicted here represent only a schematic view. The erratic path of the pole could result from variations in the dipole and nondipole components of the field or from shifts in field's main direction. The data on which the drawing of a magnetic reversal is based are from Robert S. Coe of the University of California at Santa Cruz, M. Prévot of the University of Paris and E. A. Mankinen and Charles S. Grommé of the U.S. Geological Survey.

In view of these uncertainties it is best just to summarize contemporary thinking by saying the present favorites for the main light components in the core are sulfur and oxygen. Sulfur was proposed by V. R. Murthy and H. T. Hall of the University of Minnesota primarily because it is depleted in the rest of the earth in relation to cosmic abundances. If enough sulfur could be assigned to the core, the planet as a whole might turn out to be undepleted in sulfur. It is relevant that iron sulfides are found in meteorites.

The trouble with this argument is that sulfur is relatively volatile, and it has long been recognized that the bulk earth is depleted in volatiles in relation to cosmic abundances (most notably in hydrogen and helium but also, for example, in potassium). In addition the high-pressure data indicate that only about 8 percent by weight is required for sulfur to lower the density of iron to the density of the core. That amount of sulfur is not nearly enough to make up the deficiency of the bulk earth.

This relation weakens the original justification for invoking sulfur as the lighter component, but it does not rule sulfur out. Indeed, an iron sulfide combination is regarded by many workers as being the most plausible one for the core. It is worth noting that iron sulfide is a good electrical conductor and that it melts at temperatures several hundred degrees below the melting point of mantle minerals. Its properties are therefore consistent with those of the core, and one can see how the bottom of the mantle could be solid (a silicate with a high melting point) and the outer core liquid (a sulfide with a lower melting point).

The main proponent of oxygen as the light component in the core is A. E. Ringwood of the Australian National University. He suggests that at high pressures iron oxide becomes metallic. The point is crucial because at low pressures iron oxide is not metallic. The required metallization would necessitate a drastic change of properties.

There is as yet no clear evidence by which to judge Ringwood's hypothesis, although shock-wave experiments in the laboratory have demonstrated that at something less than a million atmospheres of pressure iron oxide does undergo a transformation. Unfortunately in the experiments the nature of the transition could not be determined. In the transition does the bonding become metallic, or is there just a change in crystal structure? Answers to this question and others about the properties of iron oxide at high pressures await the conclusion of additional experiments now in progress. In any case the proposed metallization of iron oxide should not be regarded as improbable. After all, the evidence is that even oxygen can become metallic at high pressure, and molten iron oxide appears to be a semimetal at high temperatures.

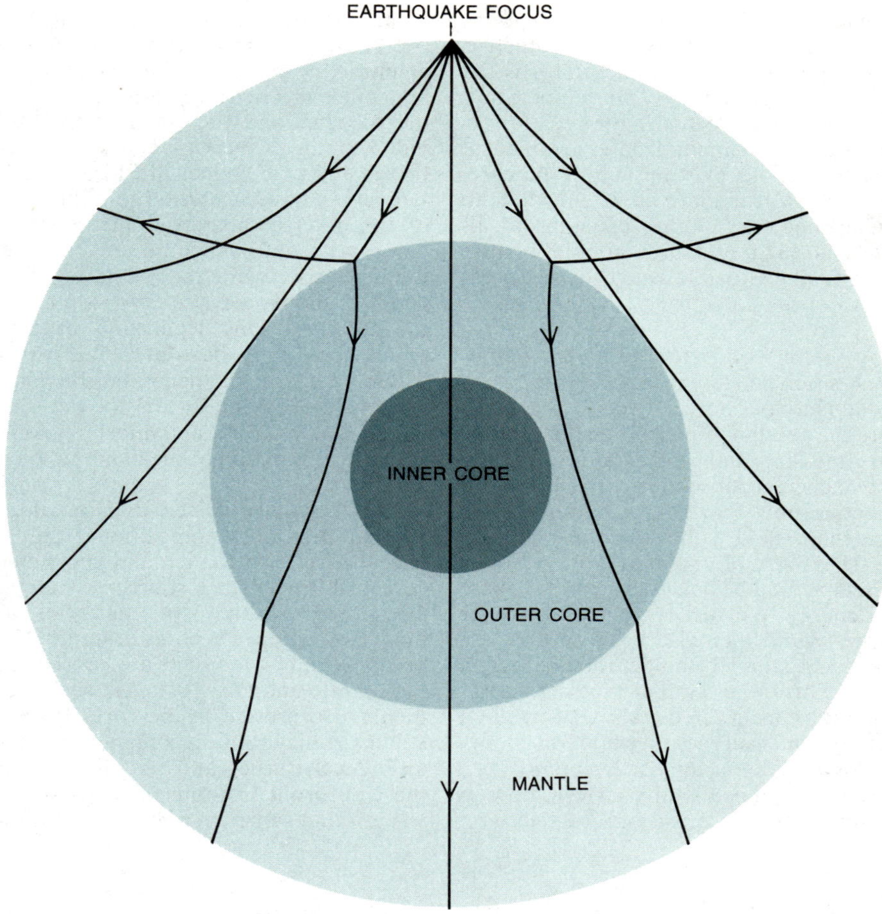

SEISMIC WAVES provide data on the physical properties of the core. Two types of waves move through the earth from the focus of an earthquake: *P*, or compressional, waves (waves that move back and forth in their direction of travel) and *S*, or shear, waves (waves that move at right angles to their direction of travel). *S* waves, which cannot travel through the body of a liquid, do not pass directly through the core, demonstrating that at least the outer core is liquid. *P* waves go through both solids and liquids. Besides information on the state of the core, seismology yields data on density, which make it possible to calculate the pressure at each depth.

One of the main differences between the oxide and the sulfide hypotheses is that under the oxide hypothesis the core must have acquired its present composition at high pressures. Ringwood concludes that below the metallization pressure oxygen does not combine with iron in any significant amount (in relation to the silicates of the mantle). This is why the core would have to form at high pressures to incorporate oxygen. In contrast, sulfur can readily be alloyed with iron at low pressures.

The effect of combining either sulfur or oxygen with iron is that the melting point of the compound is lowered. At low pressures sulfur has a much larger effect than oxygen on the melting point of iron, and it is thought this difference may persist at high pressures. Therefore a notable difference between the sulfide and oxide models for the core is that melting would begin at significantly lower temperatures in a sulfide composition than it would in an oxide one. As a result it may be easier for a core to start forming if it is sulfur-rich than if it is oxygen-rich.

In order to apply this information it is necessary to know the melting temperatures of iron alloys at the pressures of the core. The melting point of iron itself has recently been determined in the shock-wave experiments of J. M. Brown and R. G. McQueen at the Los Alamos National Laboratory. For the first time they were able to discern the onset of melting in pure iron at 2.5 million atmospheres.

Applying these data, Brown and McQueen have modeled the melting in iron alloys at pressures corresponding to those at the boundary of the inner and the outer core. On the assumption that this boundary corresponds to the melting-freezing transition and that the core is iron sulfide, they are able to estimate temperatures throughout the core. For example, they arrive at a value of 3,700 degrees C. (plus or minus 500 degrees) at the core-mantle boundary, which is close to previous estimates. This is the first time estimates of the temperature near the earth's center have been bracketed by experimental data obtained at the conditions existing within the core. Experiments that are now being done to determine the melting behavior of alloys at high pressures should further clarify the possible range of temperatures in the core.

A subtler connection between composition and temperature at the core is related to the source of energy that powers the geodynamo. Two distinct mechanisms—thermal and compositional—have been proposed for driving the convective flow that generates the earth's magnetic field within the outer core.

The thermally driven flow requires that the fluid be heated by a local source of energy. Warmer regions of the outer core would then rise because they are less dense than the colder regions, which sink. This is the familiar type of convection that occurs in the atmosphere, in a pot of water on the stove and (on a longer time scale) in the earth's mantle.

The compositionally driven flow is different in that dense and less dense regions are formed even if there are no differences in temperature. It is simply an unmixing process of the kind that can take place in a mixture of oil and water; the oil rises and the water sinks. David Gubbins of the University of Cambridge and D. E. Loper of Florida State University have suggested solidification of the outer-core fluid could result in dense crystals that would sink toward the inner core while the remaining, less dense liquid would rise toward the top of the core. This separation process could apparently be quite effective in driving the flow that generates the magnetic field. The details are not well understood, however, and most workers consider this model of the dynamo to be speculative.

The thermally driven dynamo does not call for any compositional difference between the inner and the outer core. All it needs is a source of energy. One possible source is the decay of radioactive isotopes such as uranium 238 or potassium 40, which are present in the mantle and the crust. Are such elements also present in the core? Recent studies of the uranium content of minerals by D. S. Burnett and his colleagues at the California Institute of Technology suggest that sufficient amounts of radioactive uranium are capable of partitioning into core-forming metal to be an important source of heat. Furthermore, both theoretical results and new experimental data support the idea that the chemical bonds of potassium change at high pressures in such a way that radioactive potassium could also combine with the metallic iron of the core.

In each case the data are too scanty to allow anything more than speculation about heat sources in the core. The point is that according to present knowledge radioactive decay could be the dominant source of energy driving the flow in the core. Incidentally, one reason some geophysicists argue for significant amounts of potassium in the core is that it could partly explain why the mantle and crust are depleted in potassium in relation to cosmic abundances.

There may be other sources of energy to drive convection in the outer core. For example, considerable heat could have been released when the earth originally assembled or when the core formed. This "primordial heat" hypothesis, to which I shall return, is intimately associated with whatever view one might hold on how the earth was formed.

Another possibility is that if the inner core is freezing out of the surrounding liquid, there could be enough heat from the latent heat of crystallization to power the geodynamo. This hypothesis has been discussed extensively by John Verhoogen of the University of California at Berkeley, who emphasizes the uncertainties in such a model arising from the lack of data on melting in complex alloy systems at high pressures.

Whatever the precise source of energy may be, the fundamental instability driving thermal convection is that less dense fluid lies under denser fluid. This situation arises because there is a sufficient increase, on the average, of temperature with depth and because thermal expansion causes the density of materials to decrease as the temperature is increased by the heat sources.

If the geodynamo is thermally driven, the temperature of the vigorously flowing region in the outer core increases adiabatically with depth (that is, with pressure). In an adiabatic process the energy content of a given parcel of fluid remains the same, which is to say there is not enough time for heat to flow out of the parcel as it moves over long distances, and energy is not lost to the surroundings. Thus if the parcel is compressed, its energy density increases: the fluid gets hotter. This increase in temperature with depth in the core is estimated to be rather small, about .8 degree C. per kilometer. Nevertheless, the adiabatic gradient that would exist in the thermally convecting core would have a profound influence on the evolution of the mantle and crust above.

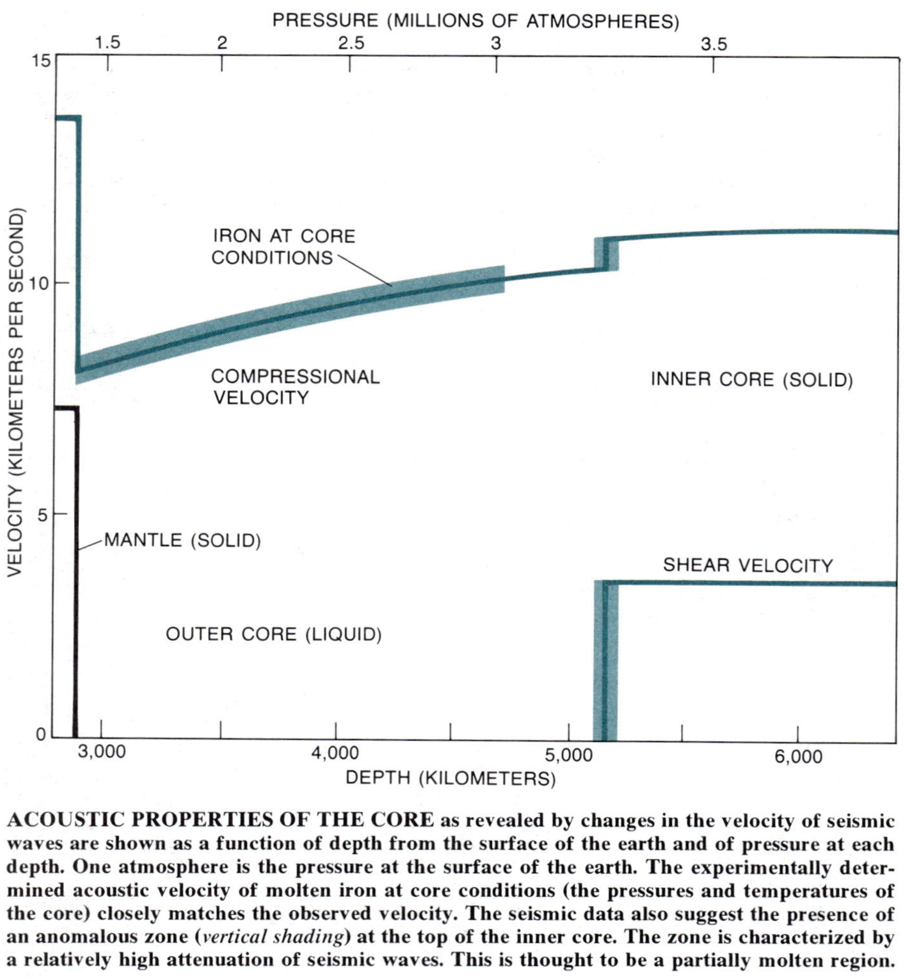

ACOUSTIC PROPERTIES OF THE CORE as revealed by changes in the velocity of seismic waves are shown as a function of depth from the surface of the earth and of pressure at each depth. One atmosphere is the pressure at the surface of the earth. The experimentally determined acoustic velocity of molten iron at core conditions (the pressures and temperatures of the core) closely matches the observed velocity. The seismic data also suggest the presence of an anomalous zone (*vertical shading*) at the top of the inner core. The zone is characterized by a relatively high attenuation of seismic waves. This is thought to be a partially molten region.

According to Fourier's law, heat is conducted down a temperature gradient with a flux (thermal energy passing through a surface of unit area in unit time) given by the value of the temperature gradient times the thermal conductivity. No one has measured the thermal conductivity of iron alloys at the temperatures and pressures of the core, but R. N. Keeler and G. Matassov have determined the electrical conductivity of alloys at core conditions by means of shock-wave experiments done at the Lawrence Livermore National Laboratory. In these metals the electrons carry the thermal energy as well as being responsible for the electrical conductivity. Thus one can employ the data to estimate the thermal conductivity of the outermost core; values between about 60 and 110 watts per degree C. per meter seem appropriate. Multiplying them by the adiabatic gradient of .8 degree per kilometer, one finds the surprising result that the predicted heat flux out of the core (70 milliwatts per square meter, plus or minus 25 milliwatts) is the same as the average heat flux at the surface of the earth.

In fact, as most investigators currently believe, if thermal convection occurs, one would expect much more heat to be transported with the fluid than conducted. Therefore the heat flux from the core into the mantle would be higher still. The core thus becomes one of the major sources of heat driving convection in the mantle. Hence there may be not only an indirect but important link between the dynamics of the core and the earth's magnetic field but also one between the dynamics of the core and the large-scale tectonic motions observed at the earth's surface. Alternatively, lower temperature gradients in the core and a lower heat flux into the mantle are possible to the extent that the dynamo is compositionally driven. A low heat flux would imply the core exerts little influence on the dynamics of the mantle.

If a thermally driven geodynamo would need no compositional difference, a compositionally driven one would need no thermal difference. Therefore if no heat sources are invoked and the convective motions in the outer core are assumed to originate in the separation of dense crystals from the core fluid, the core could theoretically be isothermal: it would show essentially no change in temperature with depth. This is an extreme case, and it seems more plausible that heat sources are at work in addition to the compositional mechanism. Hence even for the compositionally driven dynamo temperature would be expected to increase with depth in the core. The core must still be considered an important source of heat for the overlying mantle.

The compositional dynamo does have

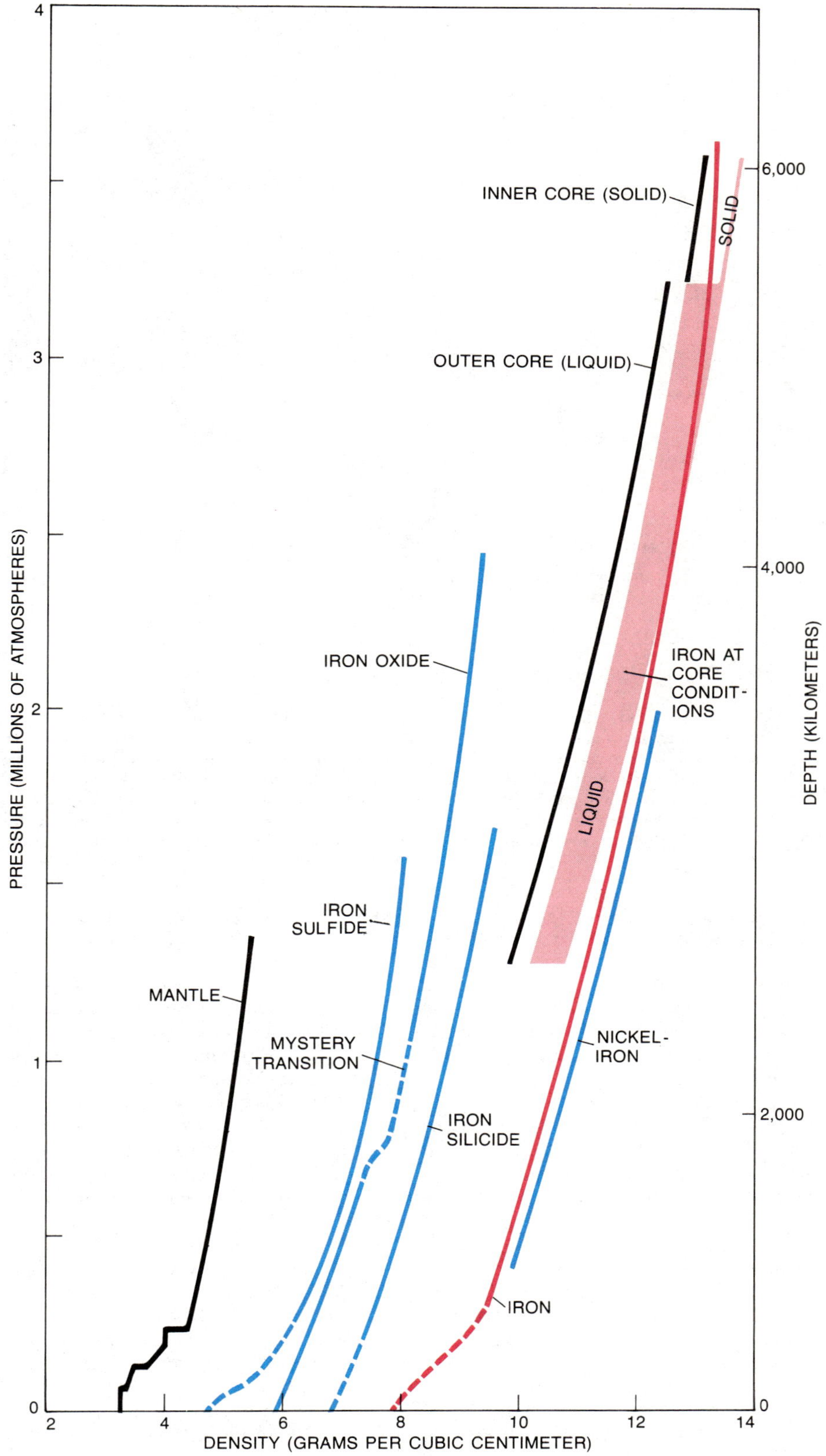

DENSITY AS A FUNCTION OF PRESSURE is charted in a comparison of the observed values for the earth's mantle and core with experimental data on iron and iron alloys that may exist in the core. The shock-wave experiments were done at the California Institute of Technology and at the Los Alamos National Laboratory. The comparison is improved by correcting the experimental data for the temperature within the earth and for the liquidity of the outer core, as is shown for iron by the shaded band. The width of the band corresponds to the uncertainty in density at each pressure. The "mystery transition" in iron oxide is a rapid increase in density with pressure that is seen in the experiments. The reason for this densification is not known, but it may indicate that the oxide becomes metallic at core pressures. If this is the case, oxygen may be the alloying element that modifies the observed density of iron in the core.

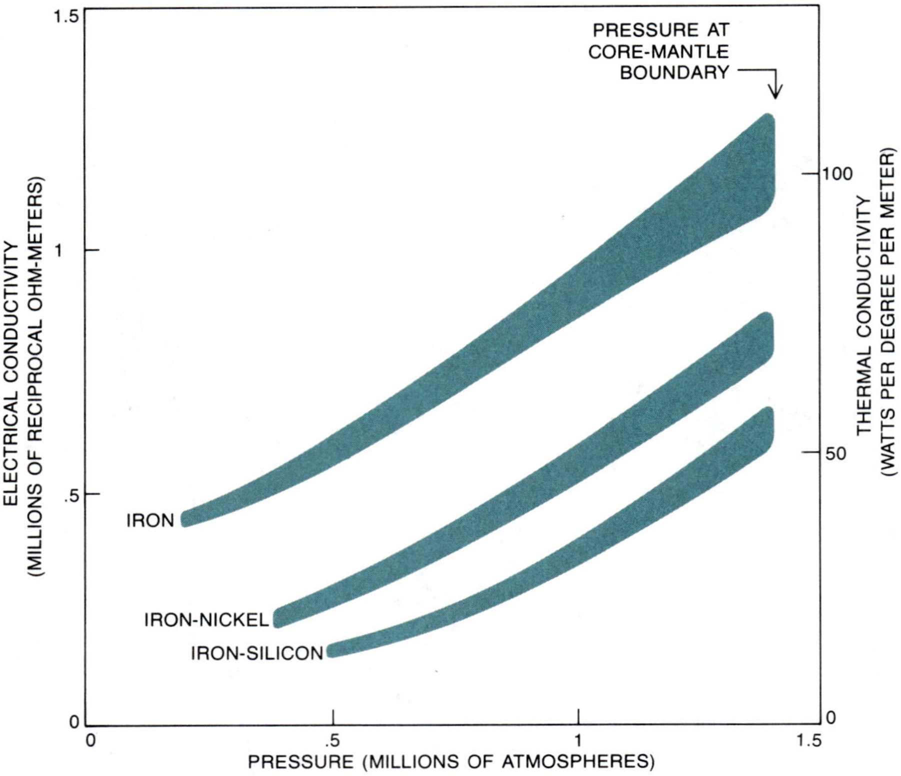

ELECTRICAL CONDUCTIVITY of iron and two iron alloys at high pressure is shown on the basis of shock-loading experiments. The data are corrected to a uniform temperature of 3,200 degrees Celsius to approximate conditions in the core. Because heat is transported by the electrons in metals, the data make it possible to estimate the thermal conductivity in the core. Near the top of the core it is evidently from 50 to 70 watts per degree per meter for the alloys, 10 times higher than in the overlying mantle. The finding suggests that the flux of heat from the core to the mantle makes the core a major source of heat driving convection in the mantle.

the simple requirement that the liquid of the outer core can separate into two phases (presumably solid and liquid) of significantly different composition, so that they can have significantly different densities. After separation the remaining liquid must be less dense than the original liquid, which makes up the rest of the outer core.

A liquid with a composition different from that of a solid is exactly what would be expected for a partially frozen alloy under equilibrium conditions. This is why an alloy melts and freezes at slightly different temperatures. Here, of course, the presumption is that the solid inner and liquid outer parts of the core are at equilibrium and that accordingly they differ in composition. Certainly the presence of a seismic attenuating zone at the top of the inner core—possible seismological evidence for a crystal-liquid mush—supports the idea.

On the other hand, one should bear in mind that the magnetization of rocks suggests the basic structure of the core (for example the presence of an inner core) may have been unchanged for most of geologic history. If the geodynamo is powered by the separation of the inner core from the outer one, it may be difficult to avoid accepting the contradictory assertion that the structure of the core has changed drastically over geologic time.

A stronger test would be to see whether the inner core is significantly denser than what would be expected for solidified outer-core material. The present seismological evidence is not sufficient to answer this question unambiguously. What seismic waves show is that the jump in density across the boundary between the inner core and the outer one is quite similar to the change in density of iron solidifying at the same pressure without a change in composition. This finding does not support the compositional-dynamo model because it implies that the solidified part of the outer core has essentially the same density as the inner core. Geophysicists are tantalizingly close to being able to evaluate this model, but more data are needed from both seismology and experiments at high pressure.

I return now to the question of primordial heat because it bears on the questions of how and when the core formed and in what way the process was related to the birth of the planet. Two extreme scenarios can be identified. One is that the earth assembled first and then separated into distinct iron and silicate fractions: the core and the mantle. The other is that the core aggregated first and then the remaining silicate-rich material was added.

The first of these pictures is called homogeneous accretion. It is mechanically analogous to the compositional model of the geodynamo, entailing a separation of dense material from less dense material after the earth had accreted. This model is well entrenched in the geophysical literature.

The second picture is heterogeneous accretion. It is a more recent and somewhat less well defined model. The reason is that different rationales have been proposed for the core metal to accrete before the silicates of the mantle do.

In order to distinguish between models of accretion one must consider the timing of three separate (but not necessarily temporally distinct) events in the earliest history of the solar system: (1) the condensation of solids out of the gaseous, cooling solar nebula; (2) the accretion of the entire earth, and (3) the accretion or formation of the core. This is the order of events for homogeneous accretion, with the key factor in the model being that all solids are condensed before accretion begins. Thus the growing planet accumulates both silicate and metal at the same time.

Subsequently the core separates from the mantle. According to Francis Birch of Harvard University, the separation leads to the release of an enormous amount of gravitational energy as the dense iron settles to the center of the planet. The amount of energy involved is comparable to the total thermal energy that would leave the earth over 4.6 billion years, given the present heat flux at the surface. It would have been enough to heat the entire planet by a few thousand degrees, which would presumably initiate substantial melting.

The best-known model of heterogeneous-accretion, proposed by Karl K. Turekian and Sydney P. Clark, Jr., of Yale University, visualizes the core material as condensing early and accreting before condensation is complete, indeed before the mantle silicates can condense and begin to accrete. Hence by the time the mantle has accreted the core is already in place because of the prior chemical separation of iron and silicates during the condensation process. The result is that the planet is relatively cold once it has formed. The reason is that little heat is retained as the earth accretes small particles condensing out of the solar nebula; most of the heat can be efficiently radiated back into space.

With no heat released by the separation of the core and the mantle, high temperatures would not be reached in this model. Yet the chemical separation of iron and silicates during condensation is no longer thought to be plausible. Calculations of the temperatures at which minerals would condense out of the cooling solar nebula indicate that the mantle phases appeared at about the same time as the core materials and maybe earlier. Therefore chemical heterogeneous accretion is now considered to be an unacceptable model.

An alternative, physical mechanism could lead to heterogeneous accretion after condensation is complete. As proposed by Egon Orowan of M.I.T. and Hitoshi Mizutani and Takafumi Matsui of the University of Tokyo, after the materials of the core and the mantle have condensed one would expect metallic (core-forming) grains to accrete preferentially. In this view, because of the brittleness and rigidity of silicate and the relative ductility and high density of iron-rich phases, one would expect a rapid agglomeration of metal and a much slower accumulation of silicates. With increasing size the growing protoplanet can accumulate silicates more readily as the gravitational attraction increases. Thus most of the mantle could be accreted after the core has assembled. In this model one can think of planetary growth as being nucleated by the formation of a metallic core.

All the evidence suggests that the earth accreted after metal and silicate particles had condensed in the solar nebula. The model of homogeneous accretion and the model of physical heterogeneous accretion could accommodate this process. The present thinking is that not only particles but also small planetesimals, perhaps already differentiated into iron (core) and silicate (mantle) regions, may have been the building blocks of the earth. One line of evidence is that iron meteorites are thought to have been present 4.6 billion years ago. They exhibit textures characteristic of a slow cooling that could have occurred only in planetary bodies with dimensions of hundreds of kilometers or more. Protoplanetary bodies of substantial size therefore could well have been present at the time the earth formed. Computer studies show that if such bodies were accreted rapidly, the radiation of heat into space would have been relatively inefficient and the planet would have got hot as it grew.

Without rejecting the possibility that the nucleation of the earth may have been initiated by the accumulation of iron, one should consider what would have happened as the planet was growing. On the assumption that both silicate and iron were being accreted, as in the homogeneous-accretion hypothesis, one finds it is not possible to delay for long the settling of iron toward the center of the planet. As G. F. Davies of Washington University has recently pointed out, the force of gravity pulls increasingly on the denser iron-rich regions with time. The reason is that the gravitational force increases as the planet grows, which is to say as its mass increases. For reasonable estimates of the sizes of the denser regions Davies finds the iron can sink, without necessarily melting, after only about one-eighth of the final mass of the earth has been accumulated. The phenomenon is explained by the fact that rocks remain relatively weak even at the pressures existing deep inside a planet.

It therefore seems inevitable that the formation of the core began well before the earth was fully formed and that the differentiation of the planet took place at the same time as most of its accretion. This picture has features of both the homogeneous- and the heterogeneous-accretion models: the earth accretes after condensation is complete, but the core is present early in the growth of the planet. One implication is that the core probably started forming at a relatively low pressure. This implication may lead to difficulties with the hypothesis that the core is an alloy of iron and oxygen. Still, the relation between the composition of the core and the processes by which it formed is not well understood, and further work in this direction is needed.

In any case, once the differentiation of iron and silicates begins the planet would be expected to heat up rapidly as gravitational energy is released. This heating and the heating caused by the relatively rapid accretion of planetesimals are thought to be enough to trigger melting, which leads to an even more effective differentiation of the planet. In this way the formation of the core is seen as a self-perpetuating and possibly accelerating process. Evidently the core had a role in triggering the geologic processes that are still taking place, some 4.6 billion years later.

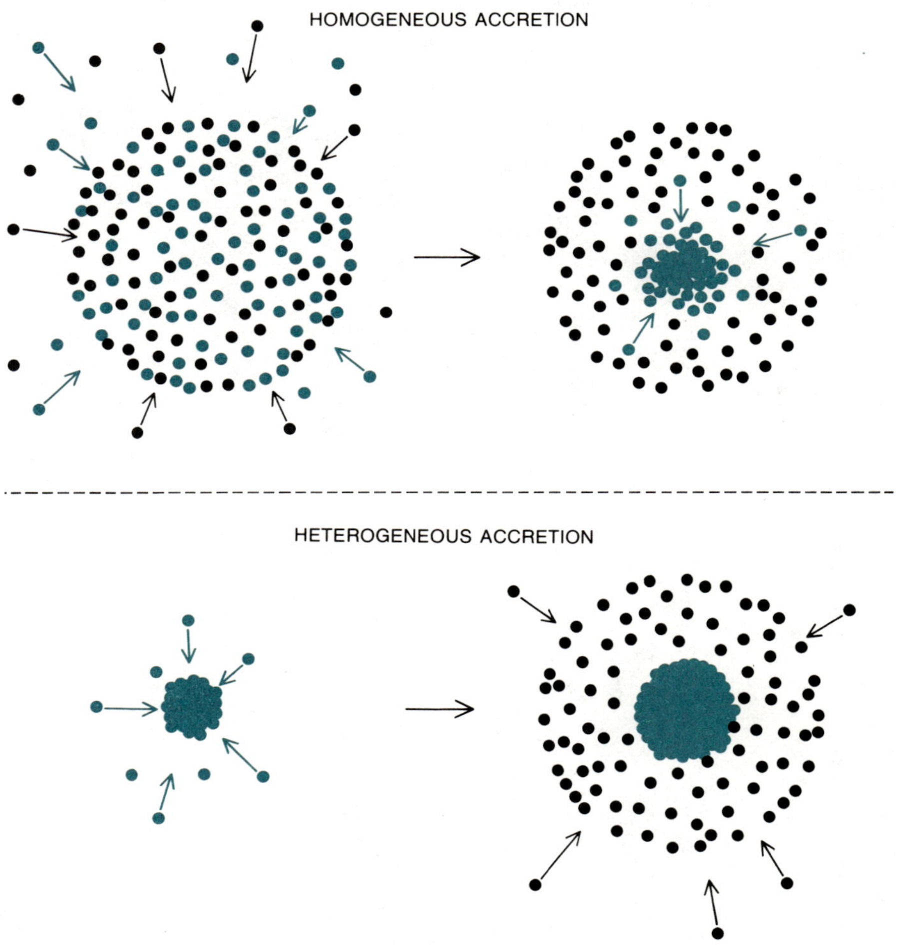

FORMATION OF THE EARTH is visualized as having been by one or the other of the processes depicted here. In the homogeneous-accretion model silicate (*black*) and iron (*color*) accumulate to form the complete planet (*top left*). Subsequently the core forms by the separation of the metal from the silicate (*top right*). During the formation of the core the iron sinks to the center of the planet and heat is generated by the release of gravitational energy. In the heterogeneous-accretion model the metallic core is accumulated first and the silicate mantle accretes around it. The sequence could occur during or after the condensation of solids out of the solar nebula, depending on whether chemical or physical processes are involved. In each model the accretion of the planet is viewed as resulting from the infall of meteoritic bodies.

3

THE EARTH'S MANTLE

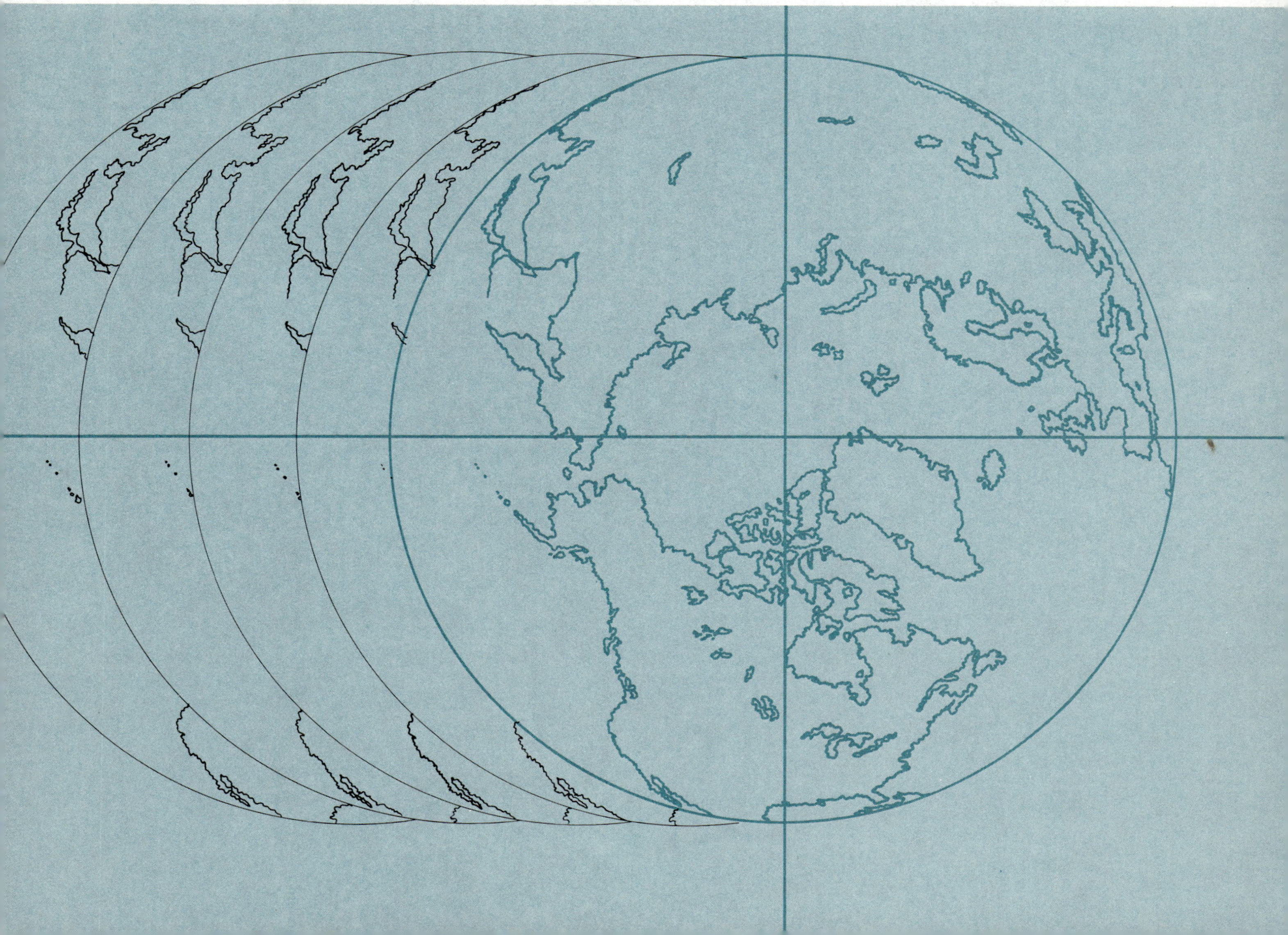

The Earth's Mantle

by D. P. McKENZIE

The great shell of silicate that lies above the metallic core is heated by the decay of radioactive isotopes. The heat energizes massive convection currents in the upper 700 kilometers of the ductile rock

Whereas the dynamics of the earth's core are reflected in the frozen polarity of ancient magnetic fields and in the luminous spectacle of the aurora, change within the mantle of the earth is recorded in the surface configuration of the oceans and the continents. Large-scale motions in the mantle take the form of currents of dense, solid, ductile material. The surface expression of these currents is the movement of the plates that form the solid surface of the earth, most of which include parts of both oceans and continents. The familiar map of the world is therefore a snapshot of dynamic processes that have their origins in the mantle. The snapshot does not, however, reveal much about the deeper circulation.

The major difference between continents and oceans is the thickness of their crust. The continental crust is anywhere from 20 to 70 kilometers thick; the mean oceanic crust is only about six kilometers thick. The plates therefore have to be at least 70 kilometers thick if they are to carry continents. It is now clear that the thickness of the plates actually has nothing to do with the thickness of the crust, and that all plates are about 100 kilometers thick, regardless of whether they are covered by continental or oceanic crust. This layer from which the plates are made is called the lithosphere.

The currents that move the plates are convection currents, in which hot material rises and cold material sinks. New plate is formed at a midocean ridge and cools as it moves away from the ridge toward an oceanic trench. The plate is then subducted by the trench and sinks into the mantle, because it is colder and denser than the surrounding material. As the dense plate sinks it releases gravitational energy and carries cold material downward, forming a sinking cold limb of a convection cell. These two parts of the convective circulation, namely subduction and the motions of the plates at the surface, can be directly observed.

In addition there must be a horizontal return flow that carries material from the trench back to the ridge, and an upward movement of hot material to balance the downward movement near trenches. Nothing can be learned about this part of the circulation from the movement of the plates. In particular there is no reason to believe the upward flow coincides with the position of a midocean ridge; indeed, it is now known it does not. The motion of the plates also tells us nothing about any mantle flow whose horizontal extent is considerably smaller than the size of the plates themselves. The reason is that the plates move rigidly, and their motion responds only to the average of the forces on their base and their boundaries. Clearly little can be learned about mantle convection from past and present plate motions, even though these motions are now known in great detail. This situation was thoroughly unsatisfactory: everyone believed convection in the mantle provided the energy to maintain the motions of the plates but no one could explain in any detail how it did so!

Fortunately the behavior of convecting fluids had been studied in considerable detail by workers interested in fluid dynamics, who had developed a variety of methods for modeling such flows. The most straightforward models consist of a layer of fluid in a shallow tank heated from below or from within. The behavior of the fluid can be studied by measuring the temperature within it or by shining light vertically through it; the light is focused by cold, sinking sheets to form bright lines. Another method of investigating convection, which Nigel O. Weiss and I have been pursuing intensively at the University of Cambridge, is to solve the relevant equations numerically on a large computer. The two methods are in many ways complementary. The tank experiments can easily be done in three dimensions, but they are little help in understanding geophysical observations. The numerical experiments are at present limited to two-dimensional flows, but they can serve to relate the observations to the convective circulation. In carrying out such modeling our aim was to understand the physical processes that control convection, not to create a mathematical model of the mantle.

Both the tank and the numerical experiments produced convection cells in which the separation between cold, sinking regions was about twice the depth of the convecting layer. Tank experiments with a moving upper boundary showed that the convection could then occur in rolls parallel to the direction of motion of the boundary. The convecting layer in the earth was thought to be about 700 kilometers thick, and so these fluid-

WINDOW ON THE EARTH'S MANTLE is afforded by the island of Hawaii, which has been built up entirely by volcanic eruptions. Two major volcanoes make up the island; Mauna Kea, in the upper part of the photograph on the opposite page, is now dormant, but Mauna Loa, in the lower part, has erupted several times in this century. Older, weathered lava radiates in brown bands from the summit of Mauna Loa. Darker radial bands mark relatively recent lava flows. The major site of current volcanic activity, however, is along the southeastern flank of Mauna Loa at the Kilauea caldera and near the coastline southeast of Kilauea. The chronological trend is therefore from the northwest to southeast; indeed, along the entire Hawaiian ridge the seamounts and islands become younger to the southeast. The oldest parts of the ridge are the Emperor seamounts in the northwest Pacific, and the youngest island is the one in the photograph. The lava emerges at the center of a hot, upwelling region in the mantle, which is much larger than the Hawaiian ridge itself. The ridge is thought to record the slow northwesterly movement of the Pacific plate with respect to the upwelling region. Isotope measurements on Hawaiian lava show that the material has not been well mixed with the rest of the upper mantle and must be derived instead from a part of the mantle that has been isolated from the convecting parts of the upper mantle for perhaps a billion years. The lower mantle is a possible source. The photograph is a digital reconstruction of two images made by Landsat; the reconstruction was done by the Flagstaff Image Processing Facility of the U.S. Geological Survey.

dynamical experiments suggested that mantle convection consisted of at least two scales of motion. A small-scale circulation with a distance of about 1,500 kilometers between cold, sinking regions should be superimposed on the larger-scale circulation that returns material from a trench to a ridge. This two-scale model of circulation in the mantle could reconcile the geophysical observations with the behavior of convecting fluids, but the existence of the small-scale circulation could not be directly observed until recently.

The key to mapping the small-scale flow came with the introduction of remote-sensing devices that have made it possible to map the earth's gravity field with great accuracy. The numerical calculations had shown that upwelling regions of convection should be associated with small, positive gravitational anomalies and that flow should push up the earth's surface. Plates are too thin to have much effect on either the gravity field or the surface deformation; therefore if the gravity field and the surface deformation can be accurately mapped, it should be possible to see through the plates and map the convective circulation under them.

In practice it is not the gravity field but the shape of the sea surface that can be measured from a satellite. Since the water is attracted to regions where the gravity field is stronger, there is a close relation between variations in gravity and the height of the sea surface. In this way Barry Parsons of the Massachusetts Institute of Technology, Anthony Watts of the Lamont-Doherty Geological Observatory of Columbia University, Micheline Roufosse of the Center for Astrophysics of the Harvard College Observatory and the Smithsonian Astrophysical Observatory and I have been able to map the small-scale convection cells in the mantle. The maps confirm the general features of the two-scale model, but they leave two major questions unresolved: How does the flow evolve, and to what depth does the circulation extend?

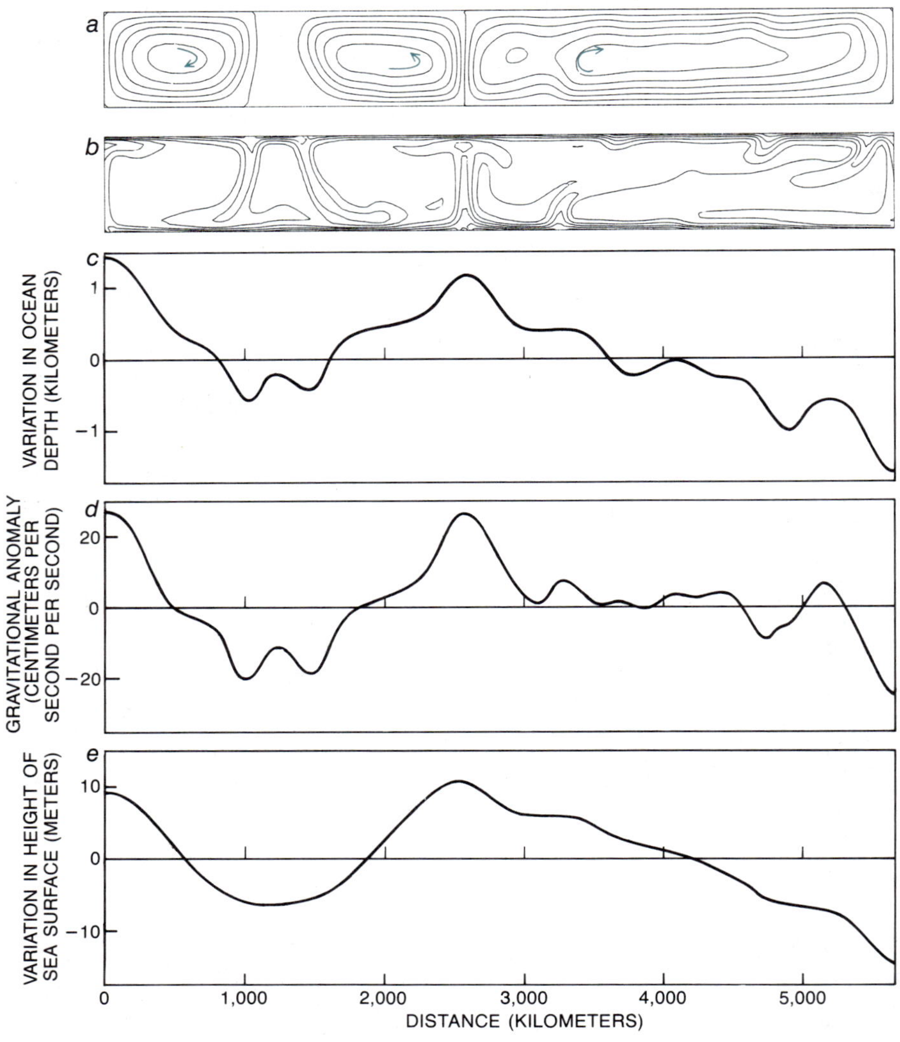

NUMERICAL SIMULATION OF CONVECTION in the mantle shows that the flow is complicated and varies with time. The simulation is carried out by assuming the mantle is a two-dimensional fluid layer of constant viscosity, uniformly heated from below, cooled from above and thermally insulated on both sides. Heat is carried through the fluid by convection; the heated material moves upward and the cooled material moves down. The circulation of the material in the cells is shown at the top (*a*); temperature contours at intervals of 100 degrees Celsius record a rising plume of heated material near the center of the fluid layer and regions of colder, sinking material at the left and right (*b*). The model can also predict certain observable effects of the convection, such as the variation in the depth of the ocean (*c*) or the variation in the local acceleration of gravity, which is termed the gravitational anomaly (*d*). An important new method for measuring the gravitational anomaly is to determine the variation in the height of the sea surface caused by the local force of gravity. The model can also predict such variation (*e*).

Before attempting to answer these two questions in detail it is worth noting they take for granted that the mantle can flow like a liquid. This behavior is quite unlike that of familiar crystalline solids at room temperature. Such solids generally deform elastically under small stresses, and they only flow when they are subjected to stresses in excess of their yield point. The yield point of mantle rocks is very large, much greater than any estimate of the stresses in the mantle, and so many geophysicists did not believe the mantle could flow. The solution to the problem became apparent when the flow behavior of materials close to their melting point was examined. Such studies were required in order to predict the behavior of materials at high temperatures, for instance those encountered in jet engines and in nuclear reactors. It was found that under these conditions all crystalline materials flow under any stress, however small it may be. The process is called creep, and it is familiar at low temperatures. High-temperature creep, however, differs in many important ways from low-temperature creep.

The creep of solid rock at high temperatures has resolved an important puzzle in geology. Observations made in the 19th century clearly showed that mountains, in spite of their great mass, are not associated with a positive gravitational anomaly. The excess mass above sea level must therefore be compensated for by low-density roots thrust into the mantle. Geologists had recently discovered that major mountain belts were raised by the thrusting of sheets of continental crust over each other, and so it was natural to argue that the same process could produce the low-density roots. The entire mass floats in the denser mantle, as icebergs do in water. The process is known as isostatic compensation. Furthermore, it was widely believed the lithosphere was everywhere underlain by a dense molten layer from which volcanic magma came, and it seemed entirely reasonable that the flow associated with isostatic compensation occurred within this layer.

Such evidence that the mantle could flow was largely neglected, however, when seismologists began to probe the structure of the mantle by analyzing the waves generated by earthquakes. They found that *S* waves, or waves oscillating at right angles to the direction in which they propagate, can travel almost everywhere in the mantle. Such waves cannot pass through the body of a liquid, and so the 19th-century idea that the lithosphere floats on a molten layer of magma had to be abandoned. It was not until the discovery of sea-floor spreading in the late 1960's and the recognition that this newly created oceanic lithosphere eventually cools and sinks once again into the mantle that the seismological evidence was reconciled with the presence of mantle flow. Indeed, if high-temperature creep had been widely recognized in 1910, when Alfred Wegener first proposed that the continents drift, the response to Wegener's ideas might have been more favorable than it was.

Ironically, some of the most convincing evidence for flow within the mantle has now been derived from seismic studies. The foci of earthquakes under island arcs define a sinking slab whose thickness is rarely greater than 50 kilometers, and that dips from the oceanic trench at the surface at an angle of between 30 and 90 degrees. The shape of the slab is controlled by the shape of the island arc, which overrides the slab and shears it like the cutting tool of a lathe. Thus the data from seismic studies show a close connection between material near the trenches and material as deep as the earthquake foci: about 700 kilometers.

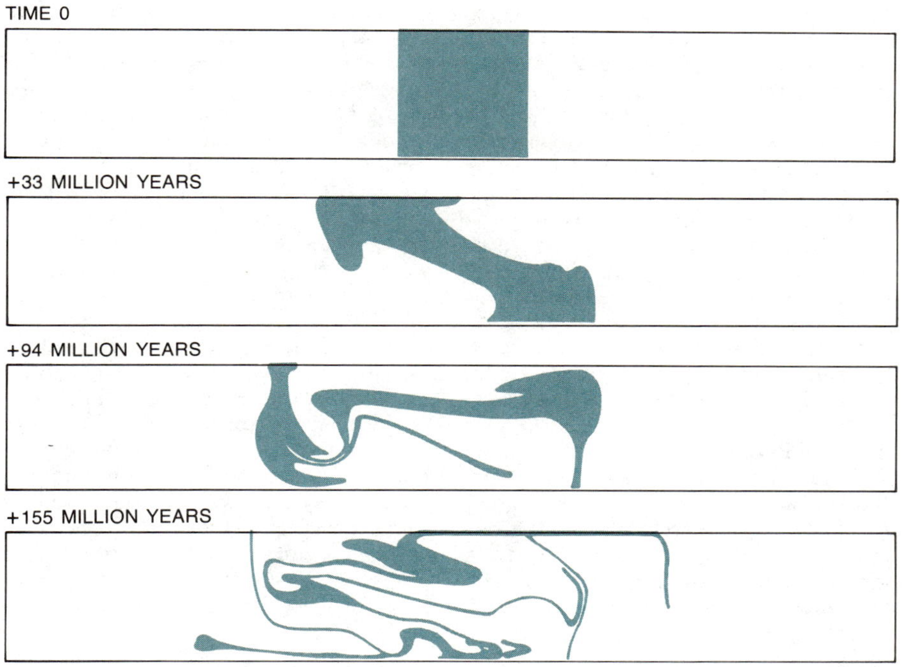

CONVECTIVE STIRRING IN THE MANTLE is vigorous enough to mix thoroughly the material in the upper mantle within several hundred million years. The stirring can therefore account for a surprising uniformity of the basalts collected along midocean ridges throughout the world: all such basalts include certain trace elements, such as neodymium, whose isotopes are found in a constant ratio to one another. Here a numerical model similar to the one on the opposite page is used to follow the evolution of a square patch of fluid whose properties are identical with those of the rest of the fluid in the convecting layer. The material in the patch is deformed into thin sheets on a time scale that is short compared with the time necessary to generate measurable variations in the isotopic ratios among the trace elements. This study of mixing in the mantle was carried out by Nicholas Hoffman of the University of Cambridge.

According to the theory of plate tectonics, the slabs are large, cold layers of mantle material that form the lower part of a plate. The slabs sink because they are denser than the hotter material that surrounds them. As they move downward and away from the trench, they tend to fracture vertically, and the part of the slab below the vertical crack tends to slip even farther downward. The sudden release of energy from the slippage generates an earthquake; this kind of earthquake is not observed, however, below 300 kilometers. On the other hand, at depths greater than 600 kilometers the part of the slab below a vertical fracture slips upward, as if the tip of the slab had met with considerable resistance to its motion. Meanwhile the entire slab is heated, and it gradually becomes hot enough to creep and so ceases to generate earthquakes. The process takes about 10 million years.

The mapping of the path of a sinking slab was one of the first empirical confirmations that convection currents in the mantle exist. As I have emphasized, the descent of a large quantity of cold material must be balanced by upward movement elsewhere in the mantle, and the transport of material from midocean ridges to trenches must be matched by deeper movements in the opposite direction. In order to understand such convective motion Nigel Weiss and I began 15 years ago to construct two-dimensional numerical models, based on the best contemporary estimates of the temperature, density and viscosity of the mantle, that we hoped would clarify the main features of convective flow.

In the model we assumed that the mantle is a two-dimensional object, rectangular in shape and 700 kilometers deep. At the bottom of the rectangle a constant flux of heat energy is introduced, representing the input of heat from the core of the earth and from the deeper part of the mantle. The sides of the rectangle are considered perfect insulators, and so heat can emerge from the rectangle only from the top, corresponding to heat lost through the earth's surface. Inside the rectangle there is a fluid whose viscosity is constant and whose density varies only with temperature. The calculation is done on a computer by dividing the rectangle into many smaller rectangular cells and then calculating the temperature, the density and other properties of the fluid in each cell at a given time. The effects of each cell on its neighbors are then extrapolated over a short time increment, and the values for the properties of the fluid in each cell are recalculated. By repeating the procedure over many increments of time the evolution of the fluid circulation throughout the rectangle can be approximated to a high accuracy.

Although the model is obviously much simpler than the real mantle, the flow within the rectangle is complex and varies with time [*see illustration on opposite page*]. Several small-scale cells are generated with variable size and shape. No cells on a scale large enough to account for the plate motions are generated in this experiment (although they were in others). Upwelling regions of hot material do not remain fixed; they shift with respect to one another at a rate equivalent to about a centimeter per year in the real mantle. The overlying lithosphere would be pushed upward above an upwelling region and pulled downward above a downwelling region by as much as a kilometer. The gravitational anomaly and the associated variation in the shape of the sea surface are calculated for each point along the upper edge of the rectangle.

The calculation of the gravitational anomaly is complicated. At first it might seem that the anomaly above a hot, rising region of a convection cell would be negative, because the density of the hot material is less than that of the cooler surrounding material. The elevation of the surface above an upwelling region, however, generates a positive anomaly because of its additional mass. The two contributions tend to cancel, but the positive contribution from the surface deformation is slightly larger than the negative one from the hot material. Hence according to the model a positive gravitational anomaly should be found over the rising region of a convection cell and a negative anomaly should be found over a sinking region. In order to detect convection cells one should look for a pattern of positive and negative gravitational anomalies that is correlated with a similar pattern of anomalies in the elevation of the sea floor.

In 1978 Parsons, Watts, Roufosse and I began to analyze all available gravitational and bathymetric data in order to detect such a correlation and thereby map the circulation of the mantle. Both kinds of data present special problems that must be overcome before useful correlations can be made. The conventional method of measuring gravitational anomalies is to measure the extension of a spring in a delicate balance. The weight of the mass at the end of the spring varies with changes in the gravitational field, and so the extension of the spring gives the anomaly. The instrument works well on land but is less accurate at sea. Indeed, although we needed an accuracy of one part per million, we were not confident that the accuracy of measurements made aboard ships was better than one part per 100,000.

Remote-sensing methods offered an alternative. Seawater is attracted to regions where the gravitational anomaly is positive and forms a measurable bulge on the surface; a trough in the sea surface is formed where the anomaly is negative. The numerical model predicts that the change in elevation from bulge to trough caused by circulation in the mantle is about 20 meters over a distance of 2,000 kilometers. Although the variation is small, the effect can readily be measured with the aid of a satellite. At least two satellites, *Geos 3*, which was launched in 1978, and *Seasat*, which operated for three months in the same year before it failed prematurely, have made measurements of the height of the sea surface with radar altimeters. If the position of the satellite is known, the time radar waves take to travel from the satellite to the sea surface and back can be measured; the height of the sea surface can thereby be determined.

In order to correlate the altimetric measurements with the changes in the shape of the sea floor the depth of the oceans must be measured as well. All detailed bathymetry must still be done by measuring the interval for sound waves to travel from a ship to the sea floor and back again. Raw bathymetric data, however, cannot be used directly to map swells and depressions in the sea floor caused by the circulation of the mantle. Short-wavelength changes in depth (changes that take place over relatively short distances) are caused by elastic forces within the plates themselves and have nothing to do with the mantle. For example, when a load is placed on the surface of a plate, the plate bends downward. The most obvious causes of plate loading are volcanic eruptions, such as the ones still building the island of Hawaii.

Watts has found that the bending of the plate caused by the weight of the volcanic rock making up the island and its submarine part is confined to within less than 200 kilometers of the load. By measuring the effects of plate loading in the vicinity of many islands and seamounts Watts has estimated that oceanic plates behave as if they were elastic plates whose thickness is 35 kilometers or less. Such a plate is too thin and flexible for its elastic response to local forces to affect any bathymetry or gravitational anomaly with a wavelength of 500 kilometers or more. Thus in our study of convection in the mantle we removed all variations in the depth of the sea and in the height of the sea surface whose wavelength is shorter than 300 kilometers.

Another important effect of the plates on the depth of the sea floor is the age of material in the plate. Midocean ridges are relatively shallow because the plate is hot and has a relatively low density. As the plate spreads away from the ridge and cools it contracts; the depth of the ocean therefore increases with the age of the plate at the bottom. By dating the sea floor on the basis of magnetic anomalies associated with reversals of the earth's magnetic field, the depth of the sea caused by the contraction of the plate can be estimated. Since the contraction is unrelated to the swells and depressions caused by circulation in the

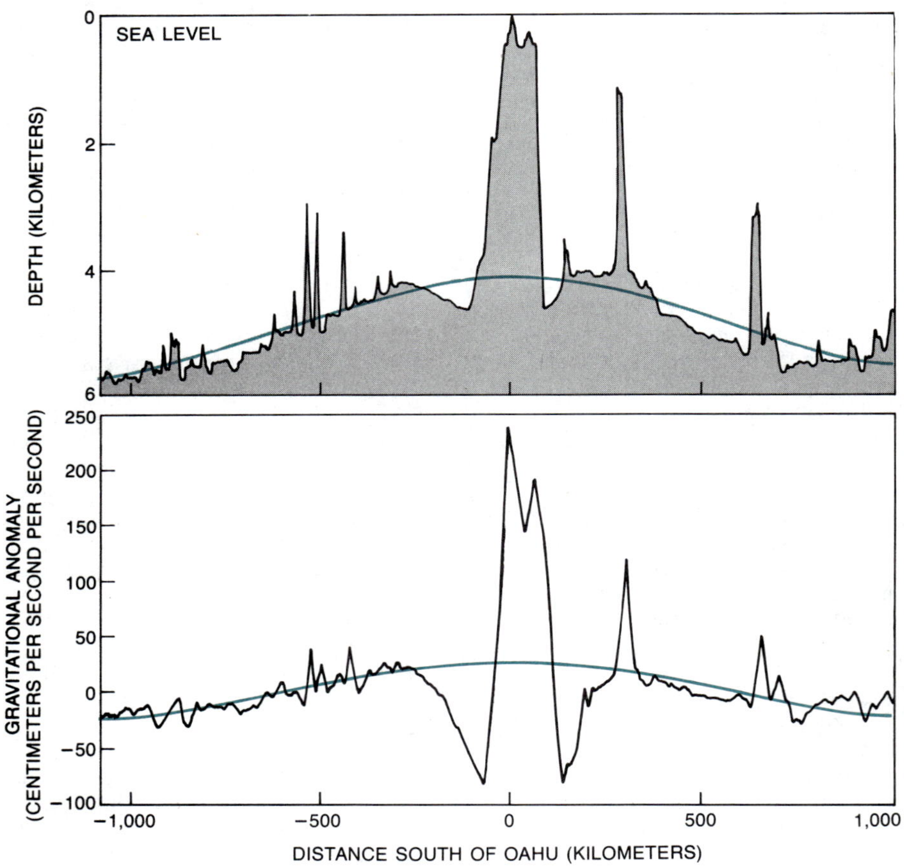

ELASTIC RESPONSE OF A PLATE to stresses that are unrelated to convection in the mantle must be understood before the effects of upwelling or downwelling in the mantle can be observed. For example, the weight of the Hawaiian ridge causes a depression of the sea floor on both flanks of the ridge, which can be seen in the profiles of both the bathymetry and the gravitational anomaly north and south of the island of Oahu. The deflection caused by the ridge is confined to within about 200 kilometers of the load. Hence it can be inferred that the rigidity of the plate is not great enough for local loads to generate variations in depth or gravity at greater distances. Both profiles also show that the Hawaiian ridge and the neighboring depressions are superposed on a broad swell, which is generally thought to be the surface expression of a hot, upwelling region in the mantle. The swell can be best observed if fluctuations in the bathymetry and gravitational anomaly shorter than 500 kilometers are smoothed (*colored curves*).

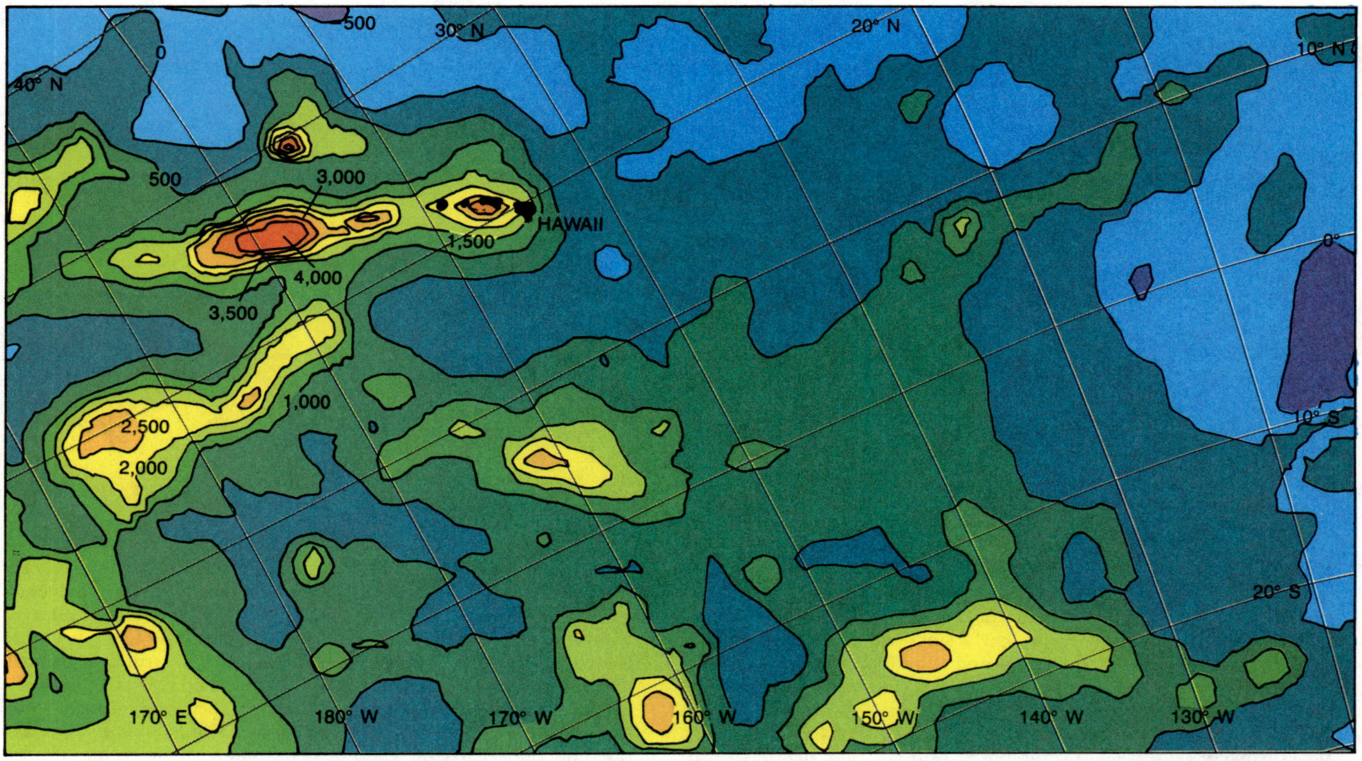

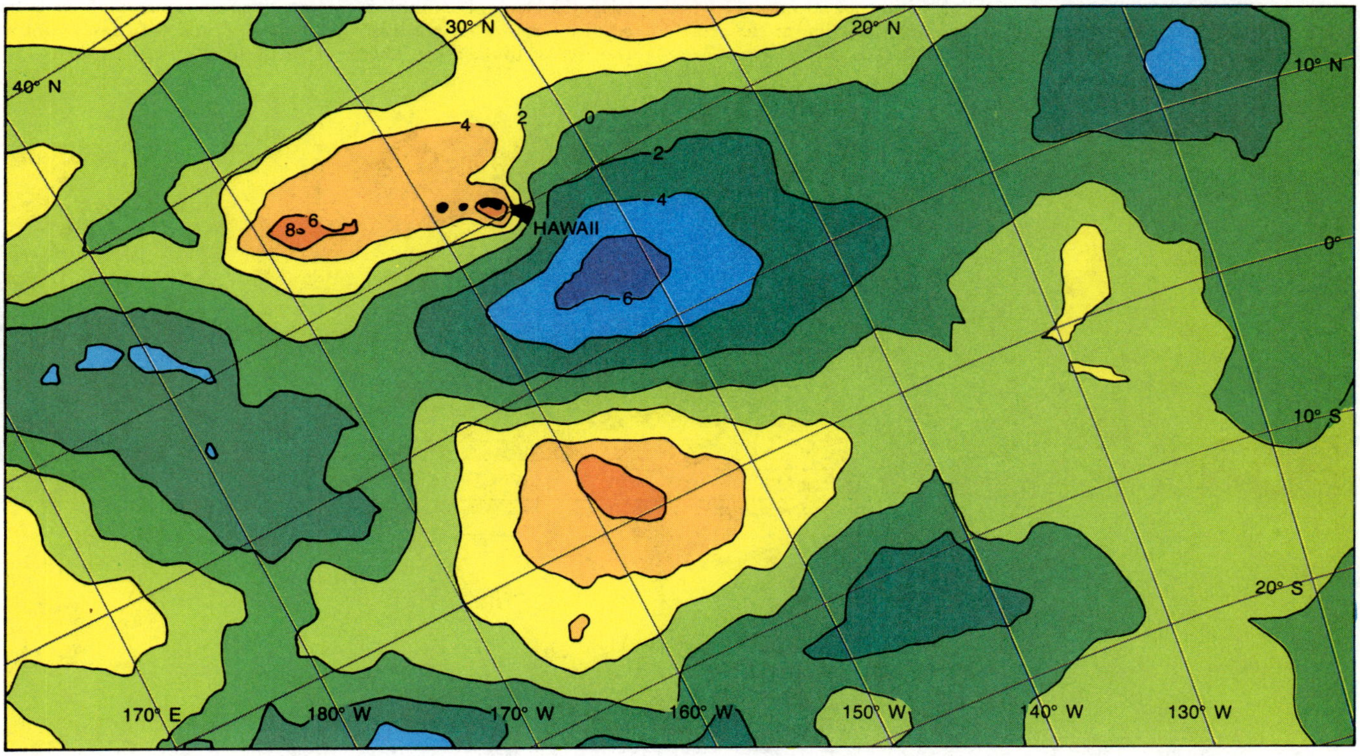

CONVECTION CURRENTS IN THE MANTLE under the Pacific plate are mapped by correlating bathymetric and gravitational anomalies. Both maps have been smoothed to eliminate fluctuations shorter than 500 kilometers; in addition the bathymetric anomaly is plotted as the so-called residual depth, which is the difference between the depth that can be attributed to the contraction of the oceanic plate as it cools and the observed depth (*upper map*). The gravitational anomaly is observed as a fluctuation in the height of the sea surface, which is measured by radar altimeters carried aboard satellites (*lower map*). The maps show that where the sea surface tends to bulge the residual depth is positive, and both features are expected above an upwelling region in the mantle. By the same token the sea surface tends to be depressed where the residual depth is negative, which is expected above a downwelling region. The map is projected in such a way that the motion of the plate with respect to the mantle is always to the left over the entire region. The motion generates a slight but detectable elongation of the anomalies in the direction of the motion, and so they resemble ellipses whose long axes run across the page.

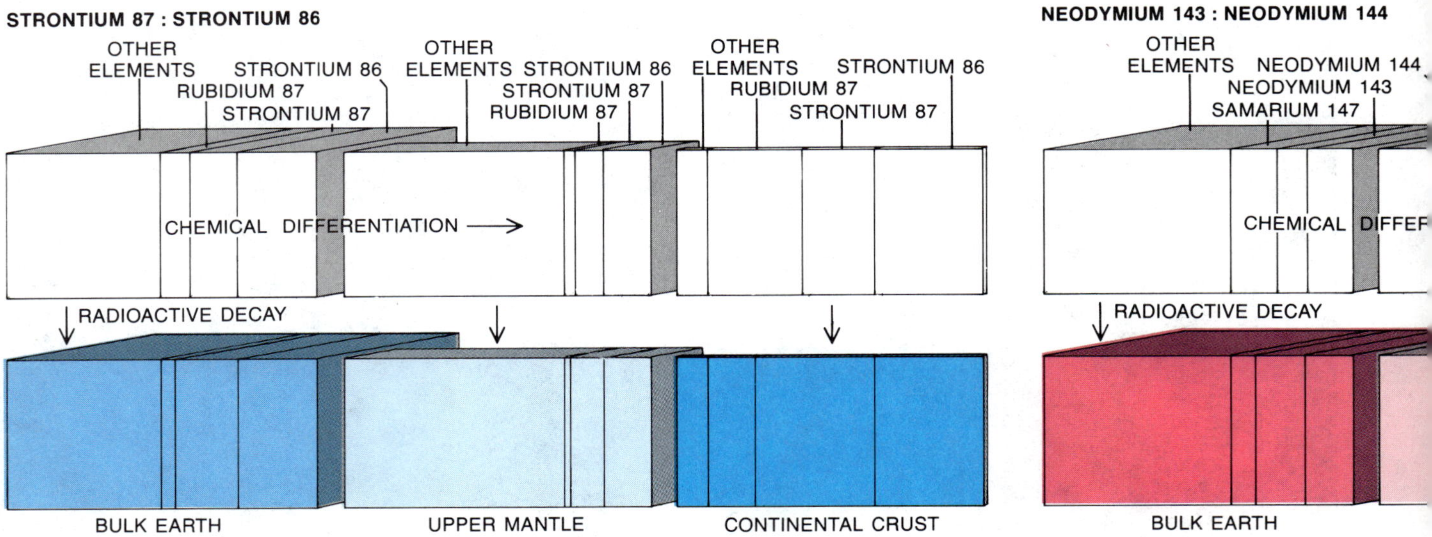

RATIOS AMONG THE ISOTOPES of an element can indelibly mark materials as having undergone chemical differentiation in the distant past, even though the ratios are not affected by recent chemical change. For example, in the early stages of the earth's history certain elements such as strontium and rubidium, whose ionic radii are relatively large, were concentrated in the crust of the earth and depleted in the upper mantle because their ions do not fit easily into the lattice framework of common minerals. The relative increase in the crustal concentration of rubidium was greater than that of strontium. In the schematic diagram the ratio of the isotope strontium 87 to strontium 86 remains the same after the differentiation of the crust and the upper mantle because all the isotopes of any element are chemically equivalent. The ratio of rubidium to strontium varies; thereafter half of the rubidium 87 in each part of the earth decays to strontium 87. (The actual fraction that decays is smaller, but here it has been exaggerated for the purpose of clarity.) Hence because of the initial surplus of rubidium 87 in the crust, the ratio of strontium 87 to strontium 86 grows most rapidly in the crust, less rapidly in the earth as a whole and least rapidly in the upper mantle. The shade of blue in the diagram indicates the present isotopic ratio: the darker the blue, the greater the ratio of strontium 87 to strontium 86. Similarly, the elements samarium and neodymium were both concentrated in the crust, but relatively more neodymium than samarium was transferred to the crust. Therefore the ratio of samarium to neodymi-

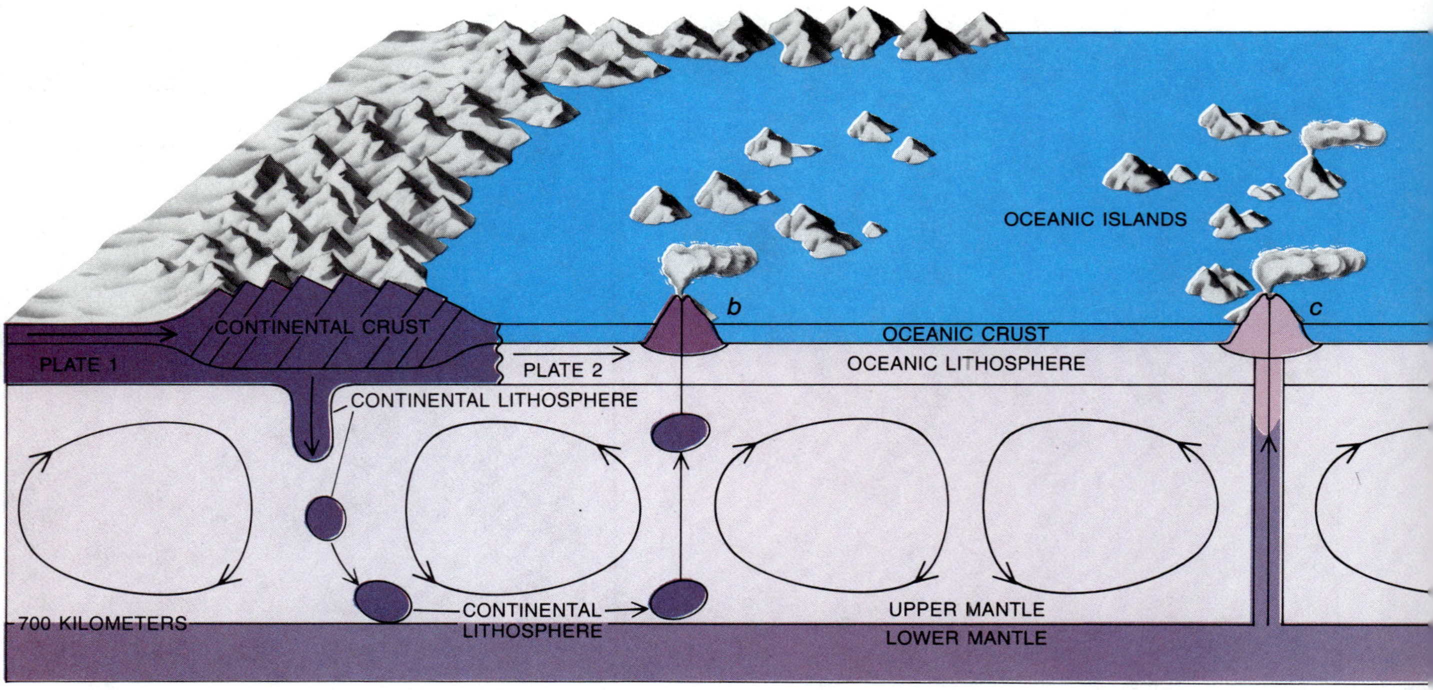

ORIGIN OF MATERIAL erupted onto volcanic islands can be inferred from the isotopic ratios of certain elements. The ratios are indicated by the color coding adopted in the illustration at the top of these two pages: material from the continental crust has a high ratio of strontium 87 to strontium 86 (*dark blue*) and a low ratio of neodymium 143 to neodymium 144 (*dark red*), and so it is colored dark purple. Similarly, material from the upper mantle is colored pale purple, and material from the lower mantle, whose isotopic ratios are assumed to be the same as those for the earth as a whole, is colored an intermediate shade of purple. Rivers carry large amounts of strontium into the oceans but only small amounts of neodymium. The ratio of strontium 87 to strontium 86 in seawater is therefore similar to that in the continents, and so it is colored dark blue. As seawater circulates through the oceanic crust it deposits continental strontium, and so the oceanic crust is also colored dark blue. Three possible origins of the material that forms volcanic islands are shown. Island-arc volcanoes are formed from a mixture of melt from the upper mantle and the oceanic crust that is subducted under another plate. They are therefore colored an intermediate shade of blue mixed with a pale shade of red (*a*). Oceanic islands associated with rising regions in the mantle have isotopic ratios of both strontium and neodymium that are between those of the upper mantle and those of the continents, and so they are colored an intermediate shade of purple (*b*). They could have been generated from a mixture of the upper mantle and

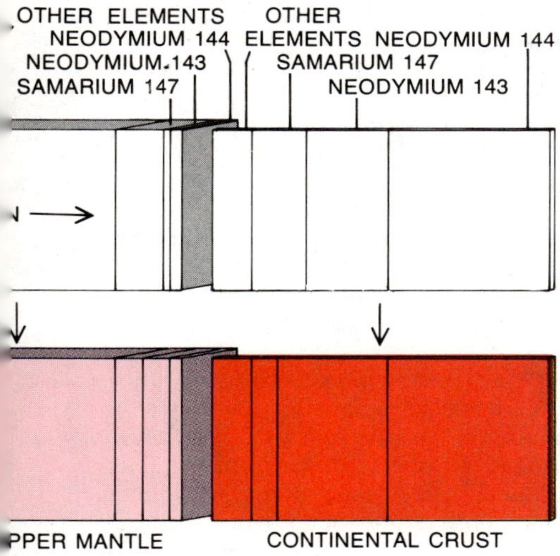

um in the upper mantle was greater than it was in the crust just after differentiation; the ratio of the isotope neodymium 143 to neodymium 144, however, remained the same throughout the earth. In the diagram half of the samarium 147 decays to neodymium 143 in each part of the earth. Therefore the ratio of neodymium 143 to neodymium 144 grows most rapidly in the upper mantle, less rapidly in the bulk earth and least rapidly in the crust. The darker the red, the smaller the present ratio of neodymium 143 to neodymium 144.

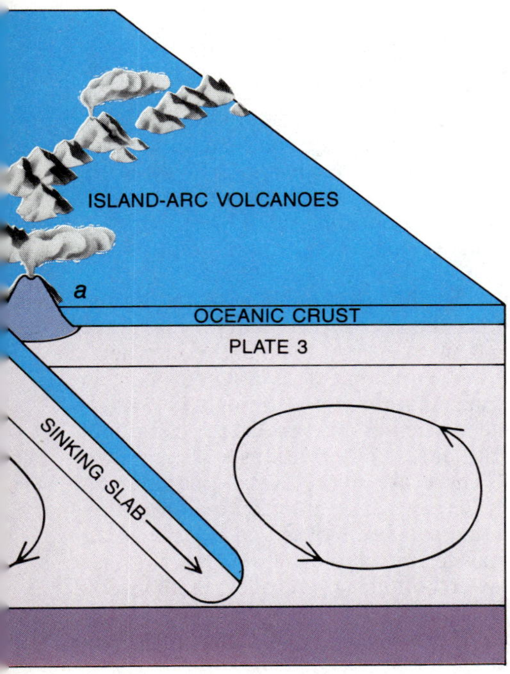

material from the mantle part of the continental lithosphere, which may have sunk into the upper mantle. Oceanic islands could also have been built up from material that leaks from the lower mantle into the upper mantle; they are colored a pale shade of purple, between that of the upper mantle and that of the lower mantle (*c*). The shades of gray of the landforms at the surface have no isotopic significance, and the vertical scale of the schematic diagram has been exaggerated. The motions of the plates are indicated by arrows.

mantle, the depth determined from the age of the plate can be subtracted from the observed depth to obtain what is called the residual depth. It is the long-wavelength variations in residual depth that we hoped could be correlated with the long-wavelength variations in the height of the sea surface.

In the Atlantic, Pacific and Indian oceans a good correspondence has been found between the two kinds of observations [*see illustration on page 31*]. For example, the sea floor around Hawaii is too shallow for its age, and so the residual depth is positive. (Both the estimate of depth from the age of the plate and the measured depth are negative numbers.) The sea surface in the same region is elevated as well, reflecting a positive gravitational anomaly. Both features are predicted to exist above a hot, rising current in the mantle. Similarly, in many regions that are too deep for their age the sea surface is depressed; the mantle under such regions is probably sinking. In all the oceans active volcanic islands are at the center of what are thought to be upwelling regions in the mantle. Perhaps the most striking feature of the observations is the absence of any obvious relation between the motions of the plates and the small-scale circulation of the underlying mantle. Upwelling regions are not in general associated with midocean ridges; they are distributed all over the ocean floor. There are even several areas along ridge axes, such as along the ridge between Australia and Antarctica, where there appears to be a downwelling current in the mantle.

The pattern of rising and sinking regions determined by this method agrees in all essentials with the predictions of the models. The size and spacing of the gravitational anomalies are similar to the calculated ones and, as was predicted, the separation between rising and sinking regions is much less than the size of the plates. The data also show a slight elongation of the pattern of rising and sinking regions in the direction of the plate motion. In other words, the large-scale circulation responsible for the motions of the plates is probably superimposed on the smaller-scale circulation that is drawn out in the direction of the plate motion. The observations therefore support a two-scale model of mantle convection.

Although our investigations of residual depths and gravitational anomalies make it possible to map rising and sinking regions, they provide little information about the depth to which the circulation extends and none at all about its evolution with time. In order to make further progress geochemical observations must be related to the known features of convective motion. Ideally one would like to be able to draw material from the mantle at selected points and match its composition with material elsewhere on the earth. In this way the mixing and distribution of the material could be traced over time. Unfortunately changes in elemental composition that are introduced as the material within the mantle comes to the surface make it difficult to infer the composition of the mantle from the composition of the melt. Progress has come, however, with advances in the design of mass spectrometers that can measure trace amounts of various isotopes.

The isotopes of an element are chemically indistinguishable: variation in the number of neutrons in the atomic nucleus has no effect on the electronic structure, which is solely responsible for the chemical behavior of the atom. Hence the ratio of the concentrations of two isotopes of a given element is not affected by the high-temperature chemical reactions that take place in volcanoes. The ratios of the relative abundances of a few isotopes do, however, change with time because certain isotopes are radioactive and decay into isotopes of other elements. Therefore any chemical reactions in the distant past that brought about changes in the relative abundances of different radioactive elements are now reflected in isotope ratios of the decay products.

For example, continental crust was made by the melting of the upper mantle, and the process concentrated in the continents more of the element rubidium than of the element strontium. If for the sake of simplicity it is assumed that the differentiation of continental crust from the mantle took place instantaneously, then just after the melting the ratio of strontium 87 to strontium 86 was the same in the continents as it was in the upper mantle. It happens that rubidium 87 is radioactive; with a half-life of 48 billion years it decays into strontium 87. Therefore as time passes the amount of strontium 87 with respect to strontium 86 increases. It does so both in the upper mantle and in the continents, but the rate of increase is higher in the continents, because after the differentiation of continental crust there is more rubidium 87 in the continents. Continental crust is thereby indelibly marked with a characteristic ratio of strontium 87 to strontium 86 higher than the ratio of the two isotopes in the upper mantle.

Isotope ratios determined by radioactive decay can also be exploited to follow the circulation of material derived from the crust, just as a colored dye added to a mass of water makes it possible to follow its movement underground. For example, it has been found that in island-arc volcanoes a certain fraction of the strontium is from continental crust and that the fraction of neodymium from continental crust in the volcanoes is much smaller. If such volcanoes were directly derived from continental crust by melting, the isotopic signatures of both strontium and neodymium in the volcanoes would be identical with those in the continents. Since the strontium shows a greater continental influence than the neodymium, there must be some process that transfers strontium from the continents to the source regions of island-arc volcanoes and leaves neodymium behind.

That process begins with continental erosion, which carries most of the strontium in the continental rocks into solution. The isotopic composition of seawater is dominated by that of the rivers, and when seawater circulates through the hot oceanic crust at the axis of mid-ocean ridges, much of the continental strontium is left behind in the basalt of oceanic crust. Most of the neodymium in continental rocks, on the other hand, does not go into solution and remains in sediments. When the oceanic crust is then carried below the island arc by a sinking plate, it melts to yield a rock enriched in continental strontium but not in continental neodymium.

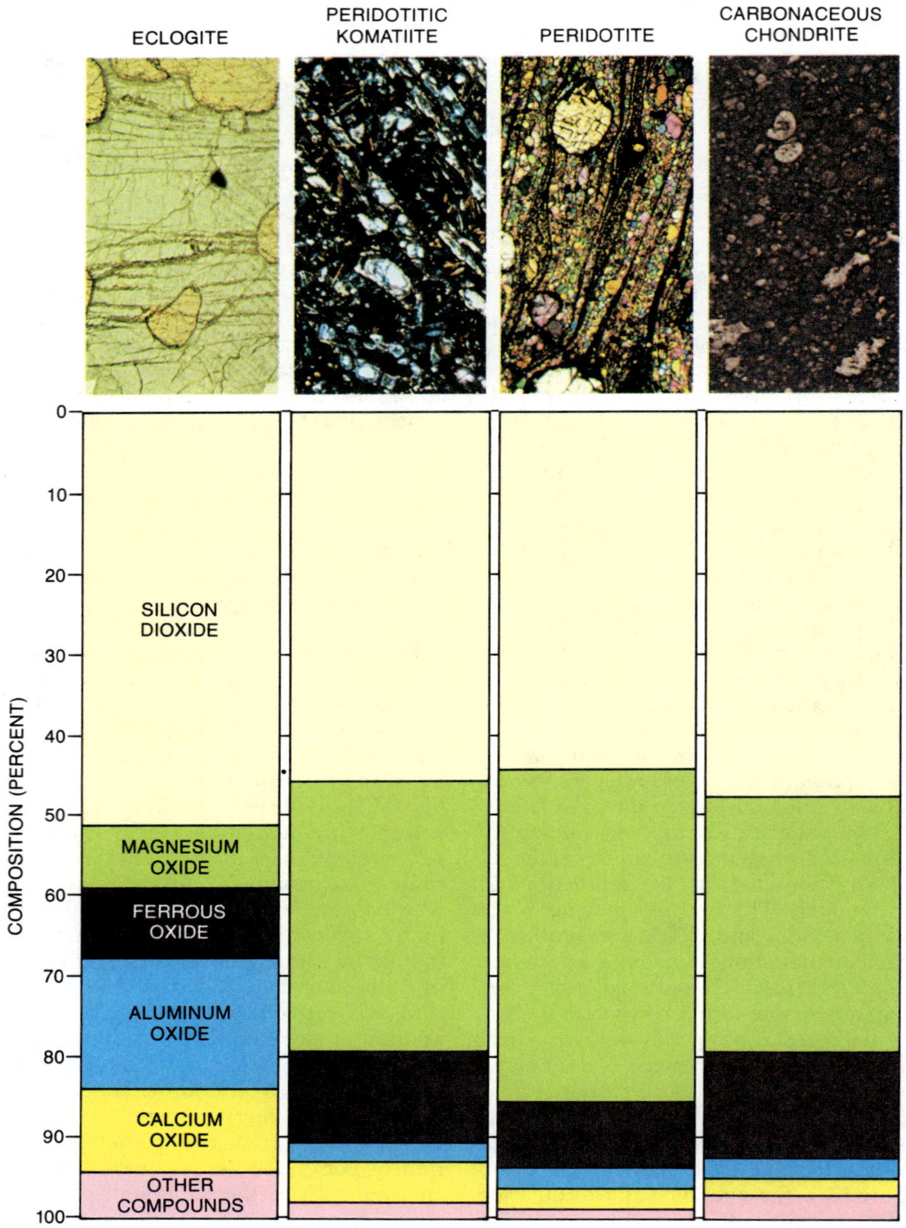

ROCK FROM WHICH THE MANTLE IS MADE must be capable of producing the melts that are known to come from it. The basalt now being erupted along midocean ridges could be formed by the complete melting of an eclogite, whose composition is shown below the photograph at the left. In the first half of the earth's history, however, melts much richer in magnesium were erupted, which cooled to form peridotitic komatiite. The long, thin crystals in this section show that this rock was once entirely molten. Such melts could have come from the rock peridotite, whose composition is much closer to that of a peridotitic komatiite than the composition of eclogite is. The composition of peridotite also resembles the oxide composition of carbonaceous chondrites, extremely old meteorites that are thought to be similar in composition to the material from which the earth was formed. Carbonaceous chondrites contain water, carbon, metallic iron and nickel and sulfides as well as oxides, but these materials have not been included in the comparison because they were largely separated from the mantle in the early differentiation of the earth. The photomicrographs of eclogite and peridotite were provided by F. R. Boyd of the Carnegie Institution of Washington, the photomicrograph of peridotitic komatiite by Euan G. Nisbet of the University of Saskatchewan and that of carbonaceous chondrite by Lawrence Grossman of the University of Chicago. The compositions given for comparison are typical, but they may differ slightly from those of the specimens illustrated.

There is a striking uniformity among all measurements of isotopic ratios made on basaltic rocks from midocean ridges, all of which show the depletion caused by the formation of continents. Such uniformity calls for vigorous mixing, and so it is natural to ask whether thermal convection could be responsible. Nicholas Hoffman of the University of Cambridge has used numerical models of mantle convection to follow the evolution of a square sample of the mantle fluid as it is deformed. He found that the mixing is vigorous, and that it can easily produce the observed uniformities in isotopic ratios within a billion years.

It is less straightforward to trace the source of the volcanic material erupted above the hot, upwelling regions of the mantle. For example, the isotopic ratios of volcanic material from Hawaii are intermediate between those found along midocean ridges and those thought to be characteristic of the earth as a whole, whereas those from Kerguelen in the southern Indian Ocean must come from a region enriched in both rubidium and samarium with respect to the bulk earth. To account for such anomalous ratios the material must have been isolated from the convecting part of the mantle for about a billion years. There are two regions that could give rise to the observed ratios: continental lithosphere and any region of the mantle, such as its lower part, that does not take part in upper-mantle convection. It is not yet clear whether either source is consistent with all the observations.

There are now several lines of evidence suggesting all scales of convection that have been observed in the upper mantle are confined to a layer whose lower boundary is no deeper than 700 kilometers. Bear in mind that the numerical simulation, which successfully predicts the distribution of rising and sinking regions for the small-scale convection cells, assumes the material circulates to a depth of no more than 700 kilometers. Isotopic ratios can also yield a rough estimate of this depth. The argument is ingenious but somewhat complicated, and it depends on knowing the total amount of the samarium and neodymium in the crust as well as in the earth as a whole.

The samarium and neodymium in the

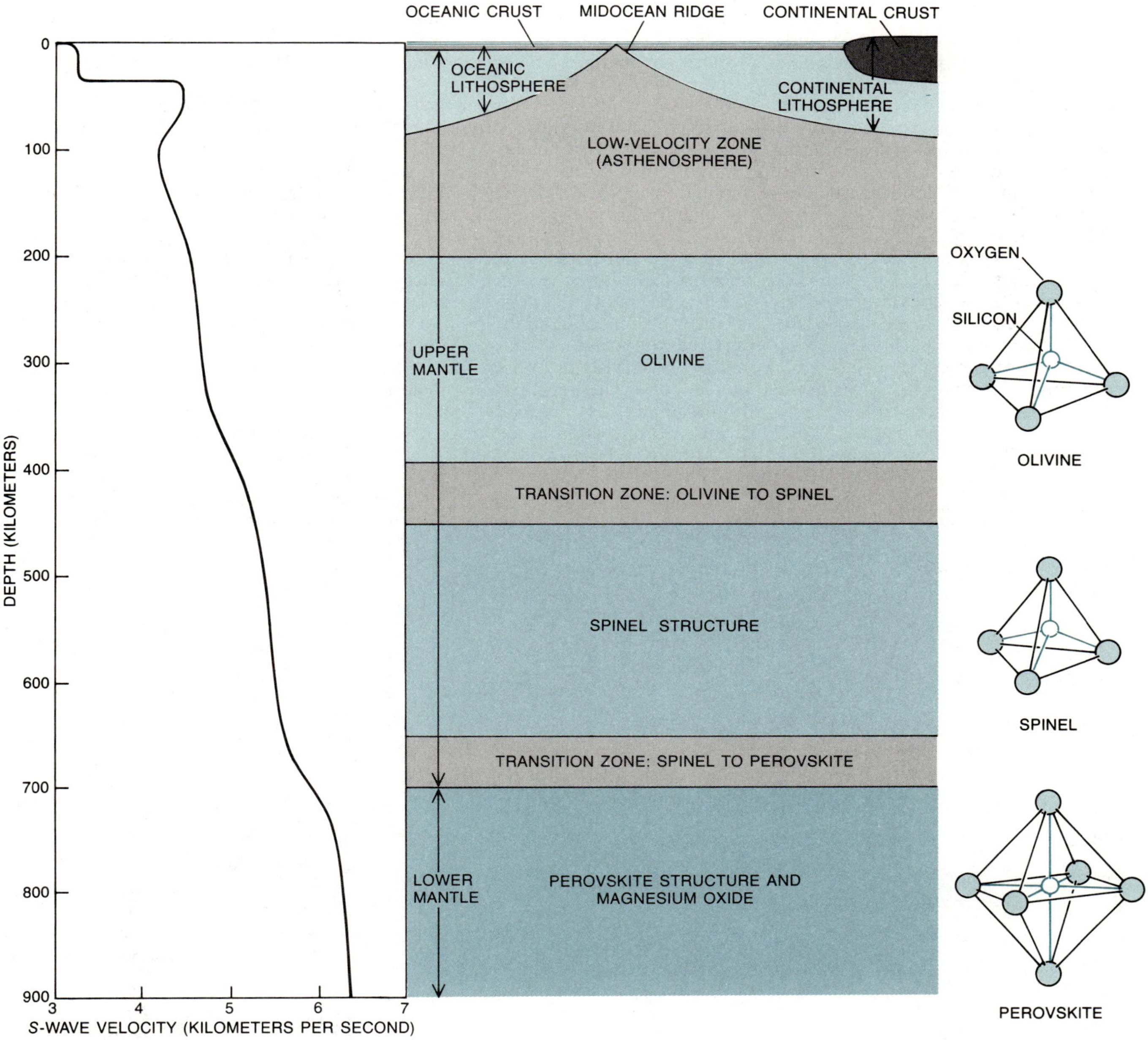

SEISMIC PROFILE OF THE UPPER MANTLE shows a series of zones in which the wave velocity is relatively constant, alternating with transition zones in which the velocity increases with depth. Laboratory simulations of the pressures within the upper mantle suggest the zones are the result of changes in the solid phase, or rearrangements of the atoms that make up the crystalline structure of the solid, rather than a result of changes in composition. The material between the base of the crust and a depth of about 390 kilometers is peridotite, whose dominant mineral is olivine. Each silicon atom in the olivine structure is surrounded by four oxygen atoms (*top right*). Below 390 kilometers is a transition zone in which the atoms in olivine are rearranged into a denser structure resembling that of the mineral spinel. Each silicon atom in the spinel structure is also surrounded by four oxygen atoms (*middle right*). The deepest major phase change is found at pressures corresponding to a depth of 700 kilometers; the spinel structure breaks down into a mixture of an even denser structure resembling that of the mineral perovskite (*bottom right*) and magnesium oxide (not shown). Each silicon atom in the perovskite structure is surrounded by six oxygen atoms. The low-velocity zone near the top of the mantle may contain a small quantity of melt.

crust can be determined from the many detailed measurements that have been made on a variety of crustal rocks. The composition of the bulk earth can be estimated from the composition of the type of meteorites called carbonaceous chondrites, which are older than the earth. The earth presumably formed out of a cloud of gas and dust or out of meteoritic bodies with a composition similar to that of carbonaceous chondrites. Although the process of formation is not well understood, it is likely that it thoroughly mixed the elemental raw materials. Therefore when the crust differentiated from the mantle, any concentration of an element in the crust would have resulted in a corresponding depletion of the element in the mantle. Since the isotopic ratios in the mantle can be measured, it is possible to calculate the volume of the mantle that had to be depleted in order to produce the observed quantity of the element in the crust.

To understand how the calculation is carried out consider two extreme cases. In the first case the isotopic ratio of neodymium 143 to neodymium 144 in the mantle is found to be infinite; in other words, there is no neodymium 144 in that part of the mantle from which the crust was produced. Such a ratio could come about only if all the neodymium in the convectively mixed part of the mantle had been removed in the course of the differentiation of the crust from the mantle. Neodymium 143 would then be generated within the depleted mantle only through the decay of the isotope samarium 147. The volume of mantle involved in crustal production would therefore have to be just enough to provide the quantity of neodymium present in the crust. Since the concentration of neodymium in carbonaceous chondrites can be measured and the total amount of neodymium in the crust can be estimated, the mass of mantle involved is easily obtained.

In the second case the ratio of neodymium 143 to neodymium 144 in the mantle is found to be nearly the same as that in the bulk earth, even though the enrichment of neodymium in the crust is the same as it was in the first case. Here the amount of neodymium extracted from the mantle by the formation of the crust must have been negligible compared with the amount left behind in the mantle. Therefore the volume of mantle involved must have been relatively large.

The real situation lies between the two extremes. The observed isotopic ratios show that the crust has been extracted from about a third of the mass of the mantle. In other words, only a third of the material in the mantle, thoroughly mixed, is needed to account for the relative abundances of the elements and their isotopes in the continents. Since a third of the mantle lies between the base of the crust and a depth of 700 kilometers, the composition of the crust and the mantle is consistent with the idea that convection in the upper 700 kilometers of the mantle is physically separated from any convection in the lower mantle. Moreover, because much of the continental crust has existed for at least two billion years, there must have been only limited transport between the upper and the lower mantle in that period.

In order to understand why the upper mantle may convect as a separate layer one must consider its composition. Whatever the mantle is made of must be capable of melting in whole or in part to produce basalt, the commonest type of volcanic rock. The entire ocean floor is covered with a layer of basalt at least three kilometers thick, which is generated by melting of the mantle under the midocean ridges. Huge volumes of basalt are also erupted by volcanoes far from the ridges. The most straightforward way to imagine making a basalt is to start with a solid rock of the same composition and melt it entirely. An appropriate rock is eclogite, which consists of garnet and pyroxene, dense minerals that are stable at high pressures.

A number of investigators still believe large parts of the mantle are made of eclogite, but there are serious difficulties with this view. In the first half of the earth's history much of the volcanic material erupted at the surface was too rich in magnesium to have been produced by the melting of an eclogite. Such a melt, however, could readily have come from peridotite, a rock whose dominant mineral is olivine, a magnesium silicate (Mg_2SiO_4). Fragments of peridotite are carried from great depths to the surface by basaltic magma, and they also dominate the rocks found in diamond-bearing kimberlite "pipes," volcanic structures that must originate in the mantle at depths of at least 150 kilometers. Most important, the composition of the carbonaceous chondrites agrees well with the composition of peridotite but not with that of eclogite.

If the composition of the mantle resembles that of peridotite, laboratory tests show there must be two changes of solid phase in the rock that forms the upper 700 kilometers of the mantle. A phase change in a solid involves a reordering of the atoms of the crystal structure in response to a change in pressure or temperature. Experiments show that the crystal structure of olivine changes to that of a spinel at pressures corresponding to depths between 390 and 450 kilometers. This change increases the density of the material by about 10 percent. The second phase change converts the spinel structure into that of a perovskite, and the density of the material increases by another 10 percent. The differences between the perovskite structure and the spinel are considerably more important than the differences between the spinel structure and the olivine. The silicon atoms in the perovskite structure are each surrounded by six oxygen atoms, whereas the silicon atoms in both the olivine structure and the

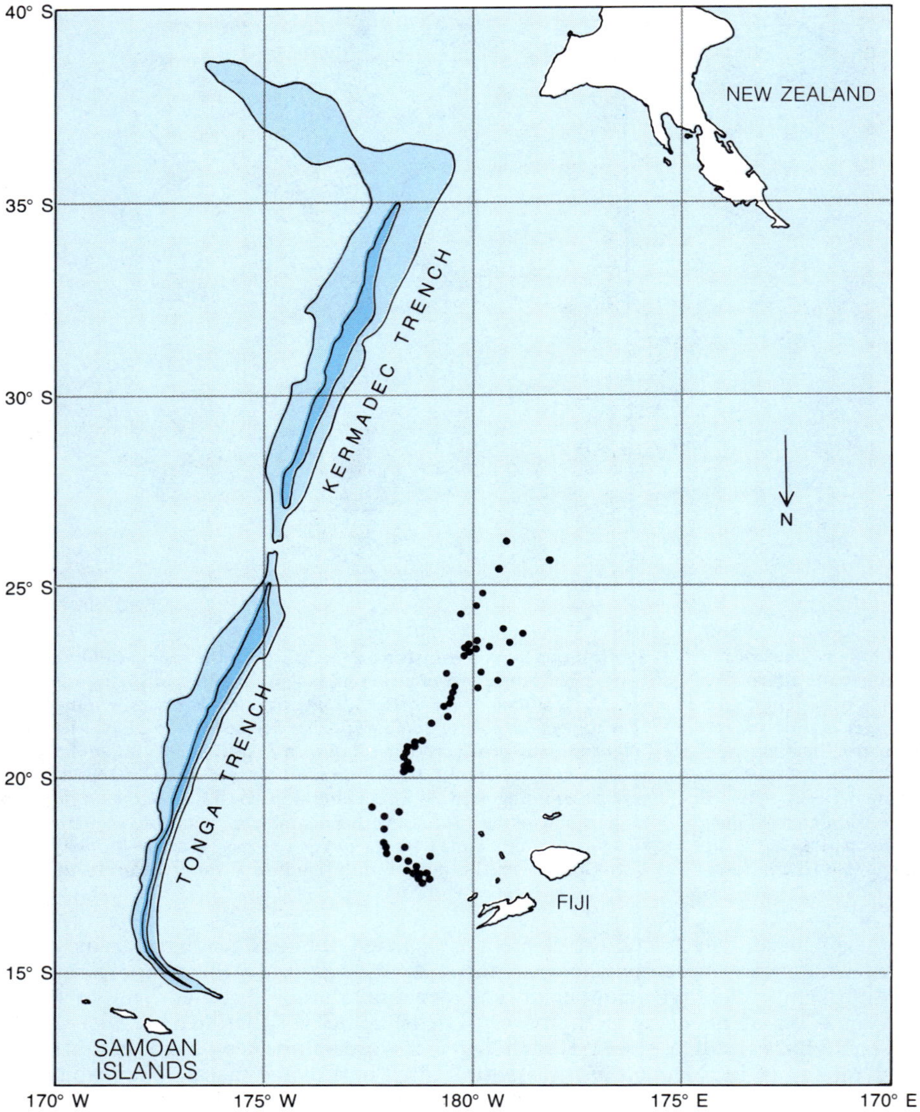

EARTHQUAKE EPICENTERS, or surface coordinates, are plotted to the west of the Tonga and Kermadec trenches in the South Pacific for all earthquakes detected at depths of from 500 to 600 kilometers below the surface. The deep earthquake foci outline a shape similar to the shape of the Tonga trench. The resemblance shows that the material at the base of the trench and the material at depth are closely connected. The diagram is based on a chart prepared by by Lynn R. Sykes of the Lamont-Doherty Geological Observatory of Columbia University.

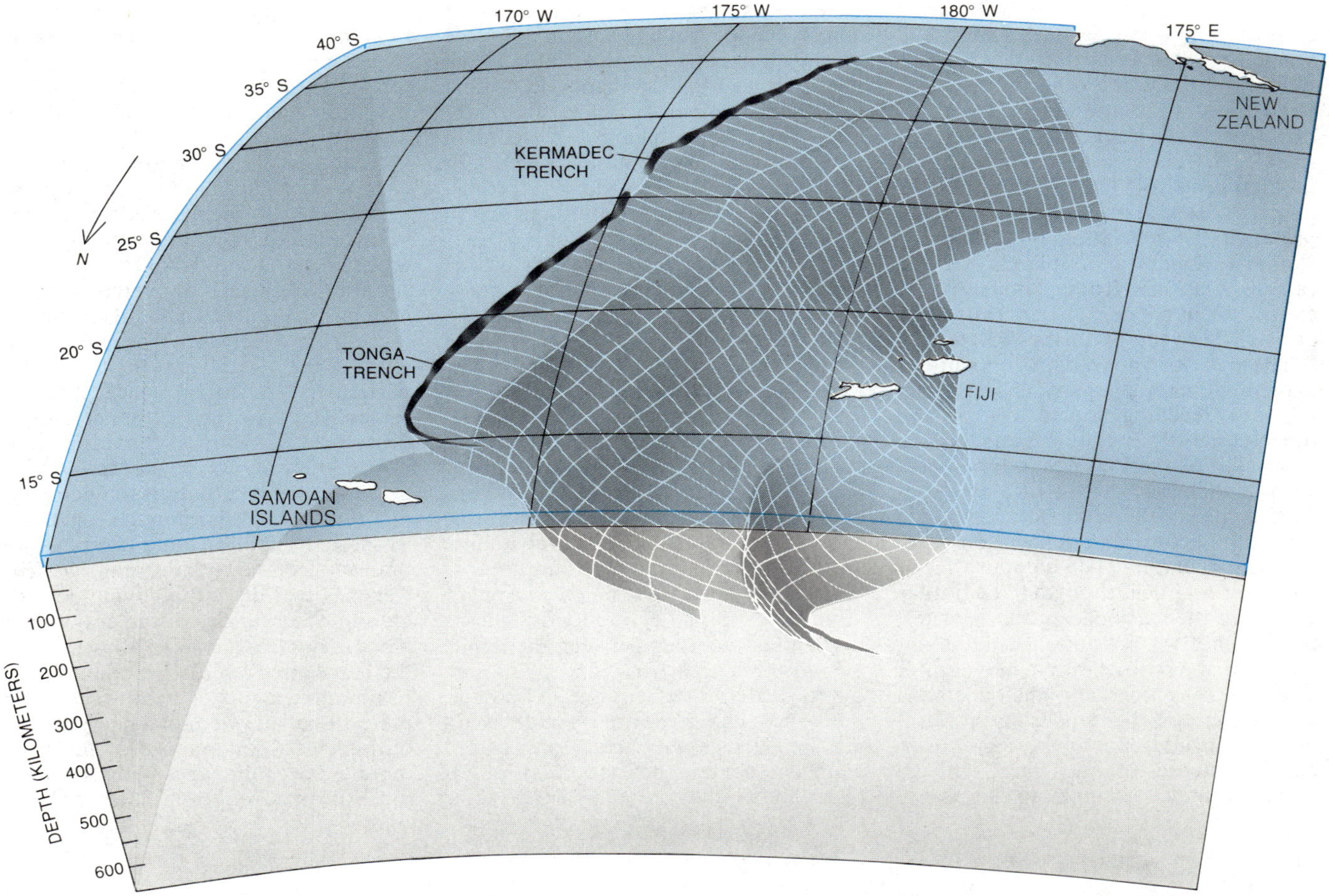

FOCI OF EARTHQUAKES at all depths to the west of the Tonga and Kermadec trenches are fitted by a computer to a smoothly descending surface. The surface can be understood as being the central plane of a descending slab of the lithosphere, which is subducted into the mantle under an island arc. The general shape of the island arc is retained in the slab at all depths. It is now thought the island arc overrides the sinking slab and shapes it like the cutting tool of a lathe. The existence of such slabs is strong evidence for relatively rapid circulation within the mantle. The diagram is based on a computer-generated surface constructed by Bryan L. Isacks of Cornell University.

spinel are each surrounded by only four oxygen atoms. The phase change to the perovskite structure takes place at the pressure that corresponds to a depth of about 700 kilometers.

The two phase changes coincide with transition zones in the mantle observed by seismologists. The velocity of seismic waves generated by earthquakes increases sharply between 390 and 450 kilometers and again near 700 kilometers. It is also striking that earthquakes have not been found deeper than 700 kilometers. Remember that such earthquakes mark the greatest depth to which a subducted slab of cold mantle can descend without becoming assimilated into the surrounding material.

The progress in the understanding of the dynamics of the mantle has encouraged a number of geophysicists to study the thermal history of the earth. Thermal evolution is regulated by the mantle because the predominantly iron core is an excellent conductor, whereas the mantle is not. In the mantle heat is transported primarily by convection. The largest source of heat loss is the process of plate creation at midocean ridges. The hot plate is cooled partly by the seawater circulating through the rock near the ridge axes and partly by thermal conduction through the sediments that cover the older parts of the plate. The total rate of heat loss is about 40 million megawatts.

Heat is generated in the earth's interior by the decay of the radioactive isotopes of uranium, thorium and potassium. The rate of such heat generation from uranium and thorium can be estimated with some confidence from the composition of carbonaceous chondrites. Potassium, however, is a volatile element, and it is less abundant in the earth than it is in carbonaceous chondrites. Its terrestrial abundance must be estimated from the composition of crustal rocks. The total heat-generation rate estimated in this way is about 20 million megawatts. Hence the earth must be cooling. There is even some direct evidence that the mantle has cooled with time. The volcanic rocks whose high magnesium content excluded eclogite as a source rock were erupted in quantity only during the first half of the earth's history. Because of their composition, at the time they were extruded they must have been about 200 degrees Celsius hotter than any volcanic rocks now being erupted.

The heat that is being lost comes partly from higher rates of radioactive heat generation in the past and probably also partly from heat generated by the formation of the earth. If the earth were a perfect conductor of heat, the rate of heat loss would always have exactly matched the rate of heat generation. The fact is, however, that the mantle is a poor conductor and can store heat, which is slowly released as it and the core cool down. The response time of the earth's heat loss to changes in its heat generation is a measure of the mantle's overall insulating properties.

Calculations have now been carried out on the assumption that convection currents carry heat throughout the mantle, and they give a response time that is too short. In other words, convection in the entire mantle would carry heat to the surface too efficiently to be consistent with the observed ratio of heat loss to heat generation. In this model the rate of heat loss could never be double the rate of heat generation because heat could not be stored in the earth for a sufficiently long time. On the other hand, if the circulation in the upper mantle is separate from the circulation in the lower mantle, the upper mantle can act as a thermal insulator. The heat stored in the lower mantle can thereby be retained. In principle the insulating mechanism is similar to domestic double glazing: the rate of heat loss from a house is decreased by introducing a convective layer of air between two panes of glass.

There is by no means unanimous agreement among geophysicists regarding the preceding conclusions or any other conclusions that depend on the properties of the lower mantle. Indeed, so little is known about the mantle at depths below 700 kilometers that there can be few constraints placed on theoretical speculation. The composition of the lower mantle is thought to be similar to that of the carbonaceous chondrites, but it is not yet possible to exclude increases in the ratio of iron to magnesium of as much as 5 percent. Such a change would increase the density by about 2 percent, in addition to the increase in density caused by the phase change from the spinel structure to the perovskite one. Even such small changes in density could have a profound influence on the interchange of material between the upper and the lower mantle. Within the lower mantle there is no evidence for any major phase changes, although it would not be surprising if some minor transition zones have so far escaped detection. Within about 200 kilometers of the core, however, the velocity of seismic waves is variable, and the waves are scattered in all directions. The cause of the variations is not clear, and no systematic attempt has yet been made to map the velocity changes and relate them to other features.

Recently a simple but ingenious device called the diamond-anvil cell has become available, which makes it possible for the first time to subject materials in the laboratory to the enormous pressures encountered in the lower mantle and in the core. A small sample of a mineral is placed between the two sharp points of gem-quality diamonds, and the material is squeezed between the points by turning a screw by hand. Although the force that generates the pressure on the sample can therefore be quite small, the area across which the force is applied is so small that the pressure between the points of the anvil can be far greater than the pressure generated by a large hydraulic press. By analyzing minute samples compressed by the anvil new phase changes have been detected that may occur in the lower mantle.

Twenty years ago many earth scientists considering the evidence for or against continental drift were consciously or unconsciously thinking in terms of a static model of the earth. This situation changed completely with the general acceptance of sea-floor spreading and plate tectonics. The effect on the study of the dynamics of the mantle was particularly profound, since plate tectonics established the existence of mantle convection without providing much information about the forces involved. Some of the first attempts to understand mantle dynamics limited the circulation to the plate motions and a return flow that carried the mantle material from trenches to ridges. The dynamic models and observations of the gravity field have now clearly shown that much of the convective circulation is not related to the movements or boundaries of the plates. I believe we now understand the outlines of the dynamics of the upper mantle; the challenge is to discover how the more massive lower mantle behaves.

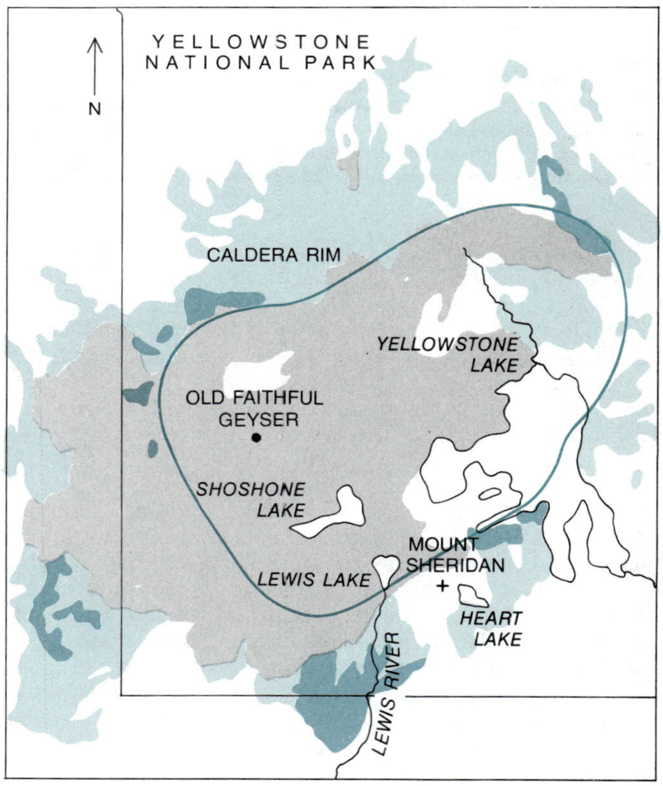

YELLOWSTONE CALDERA is one of the few volcanic structures in a continental region known to be associated with a long-wavelength gravitational anomaly and therefore with a hot, upwelling region in the mantle. In most continental areas convection in the mantle has not been mapped because the gravitational field is too poorly known. A pair of special satellites orbiting the earth at a height of less than 200 kilometers could make a worldwide map of the long-wavelength gravitational field; volcanoes that are directly related to motions in the mantle could thereby be distinguished from those that are not. Such a project is now being considered by NASA. The dark-colored areas in the map at the right are accumulations of rhyolite (a glassy volcanic rock) that began to erupt 1.2 million years ago, before the region inside the rim of the caldera collapsed; the light-colored areas are rhyolitic ash flows deposited during the climactic collapse of the roof of the magma chamber under the caldera. The gray areas mark subsequent eruptions. The satellite image was made by Landsat.

4

THE OCEANIC CRUST

The Oceanic Crust

by JEAN FRANCHETEAU

It is created and destroyed in a flow outward from midocean ridges to subduction zones, where it plunges back into the mantle. Currently it is being opened to view by submersibles and novel instrumentation

From the point of view of the earth scientist our planet probably should be called Ocean rather than Earth, not only because 70 percent of it is covered by water but also because 60 percent of its solid surface is covered by the thin crust that is manufactured in a unique geologic mill at midocean. The first piece of the oceanic crust to be identified was recovered by the British cable-laying steamship *Faraday* in 1874. The *Faraday* was sailing in the North Atlantic on a mission to mend a broken transatlantic telegraph cable. The cable had broken in 2,242 fathoms of water where it passed over a large rise in the ocean floor later named the Faraday Plateau. The *Faraday* was equipped with a large grapnel for lifting cables from the ocean floor.

According to Marshall Hall, an English geologist, "whilst engaged in grappling for the broken telegraph cable the ship caught the strong claws of the grapnel in a rock, which resisted with a strain of about 27.5 tons, to which any but a rope of marvelously perfect manufacture would have yielded. As it was, the rock gave way and a lump of black basalt came up weighing 21 lb. This mass shewed signs of having been torn off." The basaltic rock was brought back to England, and in 1876 Hall and J. Clifton Ward examined and described it.

In the century following the finding of this first piece of it the oceanic crust has come to have a central position in earth science. It is now known that the Faraday Plateau is one segment of a 59,000-kilometer system of ridges that girdle the earth under the sea. The midocean ridge has great significance in the theory of plate tectonics, which transformed the earth sciences in the late 1960's and early 1970's. The ridge marks the boundary between two rigid plates that are supported by the underlying mantle of the earth. At the ridge the plates separate slowly and the underlying rock rises to fill the gap, melting as it does so. Thus a few square kilometers of new oceanic crust is formed each year at the crest of the midocean ridges. The crust that is formed in this way is profoundly different from the crust of the continents. It is an order of magnitude younger on the geologic time scale than the continental crust and has a quite different composition from the continental masses.

As a result of many decades of observation that were unified in the plate-tectonic hypothesis it is now known where and roughly how oceanic crust is formed. The detailed structure of the crust, however, is much less well known. Therefore in the late 1970's and early 1980's the work on the oceanic crust has turned from global theory-making to investigating details of structure and composition. The oceanic crust now appears to be much more diversified both in its topography and in its layered structure than had previously been thought. The refinements in understanding the oceanic crust that have been achieved are due in large part to novel techniques of observation of the sea floor, which remains one of the least accessible parts of the planet's surface.

Satellite measurements of the gravity field over the oceans are yielding an improved picture of the general topography of the ocean bottom. The detailed topography is being mapped by new sonic methods. The layers of the crust are being probed by deep drilling, by new seismic methods and by measurements of electrical conductivity. The large body of new information that is being made available by such techniques is rapidly changing the accepted notion of the oceanic crust. The work is by no means complete, but within a few years we should have a substantially better picture of the thin crust that covers the larger part of the earth's solid surface.

In plate-tectonic theory the crust and the upper mantle of the earth are divided into the lithosphere, or strong layer, and the asthenosphere, or weak layer. The lithosphere includes the crust and part of the upper mantle. In the ocean the crust is on the average from five to seven kilometers thick; away from the ridge crest the lithosphere is about 100 kilometers thick. The lithosphere is broken up into a set of fairly rigid plates that are much like rafts floating on the less rigid material of the asthenosphere.

The plates move at a rate of a few centimeters per year with respect to each other, and the boundary between two plates can be described according to the relative plate motion. At divergent boundaries the plates separate. At convergent boundaries the plates move toward each other, and one plate generally plunges under the other and into the asthenosphere in the process called subduction. At transform boundaries the plates slide past each other. The spreading center of the midocean ridge, where molten rock from the mantle is injected into the crust, is a divergent plate boundary.

The mantle below the spreading center is composed mainly of peridotite, a type of rock consisting mainly of the mineral olivine, which in turn consists

EAST PACIFIC RISE appears as a raised area running roughly north and south in the topographic map on the opposite page showing the Pacific Ocean off the coast of South America. Brown designates shallow regions, yellow intermediate regions and green deep regions. The East Pacific Rise is part of a ridge system that circles the earth under the sea. Each ridge marks the place where two lithospheric plates diverge: the East Pacific Rise is the boundary between the Pacific plate and the Nazca plate. As the plates separate, molten material wells up to fill the gap and creates new oceanic crust. The axis of the rise is crossed by many of the large faults called transform faults. The map is based on depth soundings made by research vessels; the data were compiled by the U.S. Naval Oceanographic Office. The map was plotted and colored by a computer at the Lamont-Doherty Geological Observatory of Columbia University.

mainly of magnesium, iron, oxygen and silicon but is poor in silicon compared with the rocks of the crust. The separation of the plates at the ridge reduces the downward pressure on the mantle rock below. Hence a part of the mantle begins to move upward; the zone of upwelling extends from a depth of 50 to 70 kilometers to the base of the crust. The decompression of the mantle material is adiabatic, meaning that it takes place without loss of heat, and under such conditions the peridotite begins to melt as it rises.

Not all the peridotite melts on the way to the surface. In general the basaltic liquid injected at the spreading center is formed by the melting of 10 to 20 percent of the upwelling mantle rock. The melt collects in a magma chamber at the base of the crust, where it separates through crystal fractionation into fractions of different composition. The fractions lie above a residual solid that has the composition of peridotite. Within the chamber slow cooling and crystal fractionation result in the formation of

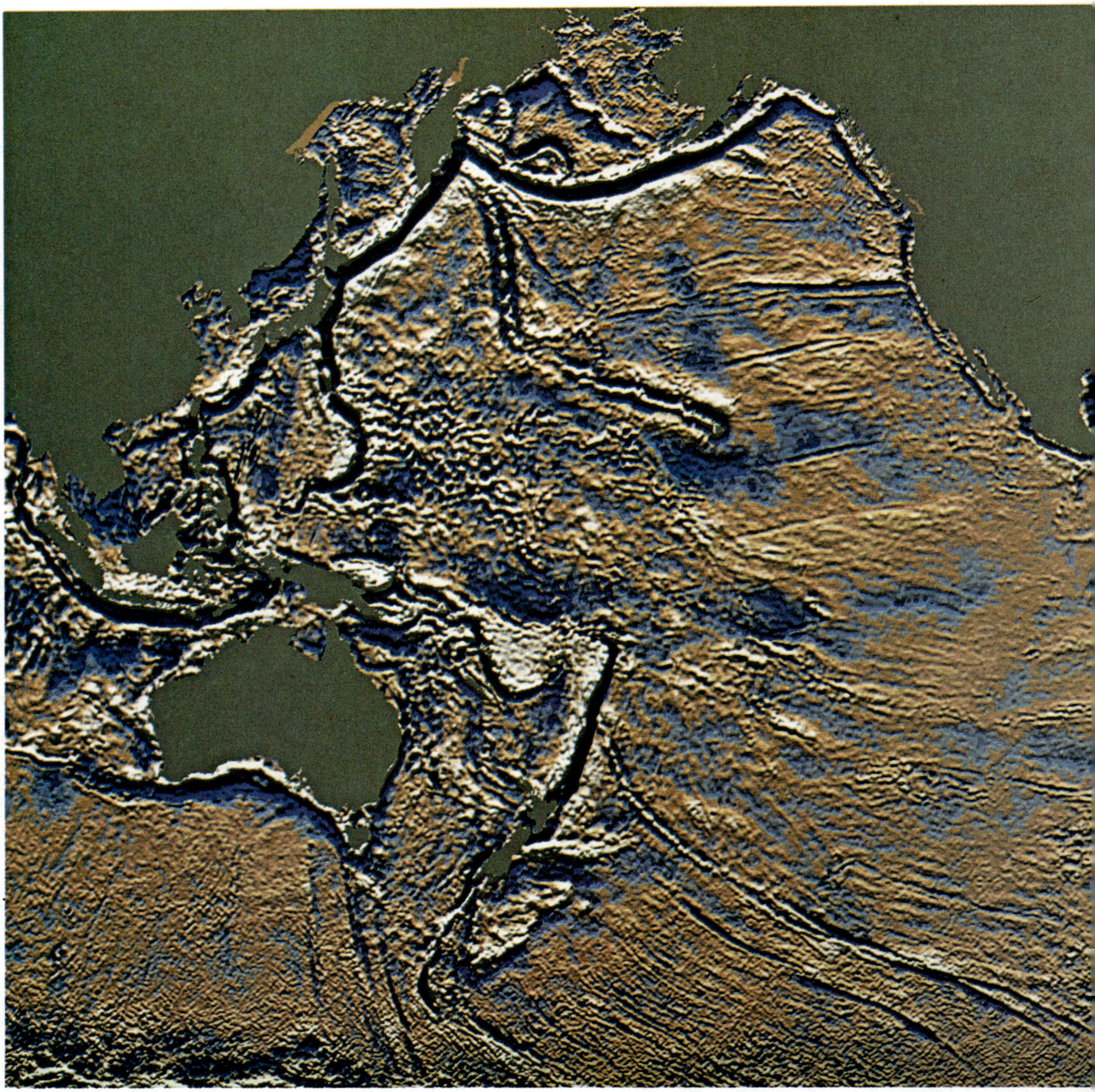

GEOTECTONIC IMAGERY based on satellite gravity measurements reveals the tectonic features of the ocean floor as if the seas had been drained. White areas correspond to the high gravity fields associated with shallow regions of the ocean floor, blue areas to the low gravity fields associated with deep regions and red areas to intermediate gravity fields. The raised red-and-white strip between Europe and America marks the Mid-Atlantic Ridge. The dark blue lines in the western Pacific show the effect of deep oceanic trenches; the trenches are surrounded by shallow areas. The irregular diagonal line in the Pacific is the Emperor-and-Hawaii seamount chain. A gabbro, a type of rock incorporating in addition to olivine the mineral plagioclase, which consists mainly of silicon, oxygen, sodium and calcium. Gabbros and other rocks that are formed by accumulation make up the lower layer of the crust. The basaltic liquid in the upper part of the chamber reaches the surface through a system of vertical passages. Once at the surface, the liquid flows down the slopes of the ridge and hardens into sheets or the rounded forms known as pillow lavas. Which of the forms results depends on the slope of the ridge and the rate at which the lava is extruded. Moreover, as the plates move apart the ascending magma hardens into a series of dikes: massive vertical sheets.

The injection of magma, or molten silicate liquid, plugs the gap left by the moving apart of the plates. The plates continue to diverge, however, and the plug is ultimately rifted open. A new cycle of asthenospheric upwelling, peridotite melting, separation in the magma chamber and extrusion begins. Meanwhile the crust formed in the previous

seamount is an undersea volcano; Hawaii is at the southern end of the chain. The map is based on measurements of the height of the sea surface made by *Seasat,* a satellite launched by the National Aeronautics and Space Administration in 1978. The computer methods for making geotectonic images were developed by William F. Haxby of Lamont-Doherty. Differences in the mass of the rock on the sea floor result in variations in gravity: seamounts have a high gravity field and trenches have a low field. Hence water tends to "pile up" over seamounts and to do the opposite over trenches. Therefore the ocean bottom can be inferred from the height of the sea surface.

round of upwelling is moving outward from the spreading center. As the crust moves outward it is modified. Tension exerted by the continuing plate motion can result in a series of fissures and faults running parallel to the strike, or longitudinal axis, of the ridge crest.

As the crust cools many small cracks appear in its upper layers. Seawater penetrates the cracks and the fissures caused by the tension in the crust. The water flows downward into the crust, is heated and rises to the surface of the crust again. Such hydrothermal circulation efficiently leaches water-soluble compounds out of the rock. Metallic elements, which readily form ionic complexes, are leached with particular efficiency. The leached elements are carried upward and the hot seawater with its burden of metals is expelled into the ocean by vents near the ridge crest. The discovery of the vents and the exotic biological communities that cluster around them was one of the most exciting findings of oceanic science in the late 1970's.

The hot circulating water and shallow heat sources lead to metamorphic changes in the lower crust and rapid chemical changes in the upper crust. In addition, as the crust moves outward a layer of sediment is deposited on its upper surface. The sediments consist largely of the remains of tiny oceanic plants and animals. The type of sediment that is deposited and the depth of the sediment cover therefore depend largely on the biological activity of the ocean.

Thus at some distance from the ridge crest the oceanic crust has the following vertical structure from the ocean floor downward. At the top is a layer of sediment perhaps .5 kilometer thick. Under the sediment is a layer known as the oceanic basement made up of interspersed sheet flows and pillow lavas with the underlying complex of vertical dikes. The basement can be two kilometers thick and is heavily fractured and altered by hydrothermal circulation. The third layer, the oceanic layer, is made up of the gabbros that solidify out of the basaltic melt in the magma chamber. The gabbros can undergo considerable metamorphism as they move away from the ridge. The oceanic layer is perhaps five kilometers thick.

This model of the layered structure of the oceanic crust and how it is formed is based on marine geophysical data, on studies of rocks from the ocean floor, on observations of fragments of oceanic crust emplaced in mountain belts on dry land and on conjectures. The rest of this article will be devoted to examining some of the unsolved problems presented by that model. It will become clear that the simple picture of the crust that was thought to be accurate until quite recently is rapidly being made more complex by new findings.

Since the first deep-sea soundings were made in the second half of the 19th century it has been known that the ocean floor lies much deeper below the sea surface than the continents rise above it. Away from the continental margins, which are not made up of oceanic crust, the ocean is on the average 3.7 kilometers deep. The great depth of the ocean and the sediment cover on the oceanic crust make the crust difficult to observe with most geologic techniques. In the past decade, however, several technical advances have greatly increased the accumulated knowledge of the oceanic crust.

At the most general level the methods of observation can be put in two groups: those that reveal the topography of the surface of the crust and those that penetrate below the ocean floor to gain information about the composition and vertical structure of the crust. In the studies of topography much interest is focused on the midocean ridge that marks the spreading center. The ridge is a long linear upswelling with a gradual slope. The average depth at the crest of the ridge is about 2.5 kilometers, and the ocean bottom slopes away on both sides to a depth of from five to six kilometers.

As the oceanic crust is carried away from the ridge crest it cools and contracts. In this process the lithospheric plate can be thought of as "floating" on the asthenosphere. Consider a block of wood floating in a barrel of water. If the block is not perturbed, it will come to rest in the water at a level corresponding to a balance between the downward force of gravity and the buoyancy of the wood. The buoyancy and hence the point of equilibrium depend on the density of the wood. Similarly, in the absence of perturbing factors the lithospheric plate floats on the asthenosphere at a depth corresponding to what is called isostatic equilibrium: the level at which the weight of the lithosphere is balanced by the pressure from the mantle. The level to which the plate sinks depends on the density of the rock in the lithospheric column, a thin vertical section through the lithosphere.

As the oceanic crust cools and contracts its density increases and hence it sinks deeper into the asthenosphere. The depth to which the crust sinks has been shown to vary with the square root of its age. Crust that is two million years old lies at about three kilometers, crust that is 20 million years old lies at four kilometers and crust that is 50 million years old lies at five kilometers. Charting the topography of the ocean bottom can therefore provide an estimate of the age of the crust.

Topographic maps can yield other significant information on the movement of the plates. Because lithospheric plates are rigid bodies, when two plates separate, their motion can be described in relation to a point on the earth's surface that is referred to as the pole of rotation. (The pole of rotation should not be confused with the earth's geographic or magnetic poles; it is significant only in relation to plate motion. Furthermore, the pole describing the relative motion of a pair of plates can shift several times over the history of the plate's interaction.) Continuous segments of spreading axis along the ridge crest define great circles that pass through the pole of rotation much as a meridian, or line of longitude, passes through the geographic pole.

As two plates move around their pole

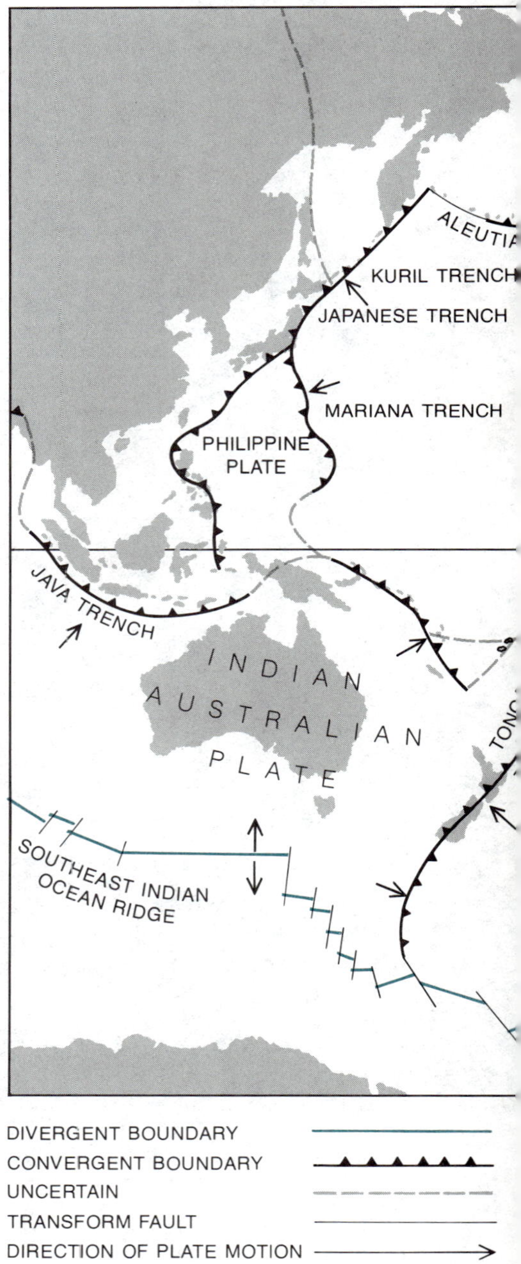

of rotation transform faults caused by stresses in the plates form across the strike of the ridge. Other transform faults are caused by the irregularity of the initial rupture within a continent that created the ocean basin. When the land masses separate, the breaks parallel to the direction of plate motion develop into transform faults; the breaks perpendicular to the plate motion develop into spreading centers. At a transform fault the axis of the ridge is offset, and the parts of the plates slide by each other in opposite directions.

The offsets at the transform faults are roughly parallel to the direction of relative plate motion. The offsets have the effect of keeping the axis perpendicular to the direction of spreading. On the Mid-Atlantic Ridge transform faults can be as close together as 50 kilometers. Away from the ridge, outside the offset, the parts of the plate are not in motion with respect to one another. In this region structures called fracture zones mark the position of the transform faults; the fracture zones extend from the ridge crest like ribs from a spine. Between the transform faults are many smaller faults that are also caused by plate motion.

The precise position of the faults and fracture zones on the ridge holds an excellent record of the kinematics of the plates: the history of the relative plate motions. Mapping techniques that enable geologists to identify the position of these transverse features can therefore serve as the basis for reconstructing the history of the plates, by a method analogous to projecting a reel of motion-picture film in reverse.

One of the most promising tools for reconstructing the kinematics of the plates is a kind of mapping that has been named geotectonic imagery by its developer, William F. Haxby of the Lamont-Doherty Geological Observatory of Columbia University. Geotectonic images come from data gathered by *Seasat,* a satellite launched by the National Aeronautics and Space Administration in June, 1978. *Seasat* was equipped with a radar altimeter capable of measuring the height of the sea surface to within five to 10 centimeters. The instruments on board the satellite failed prematurely after three months of operation, but by then the device, orbiting at an alti-

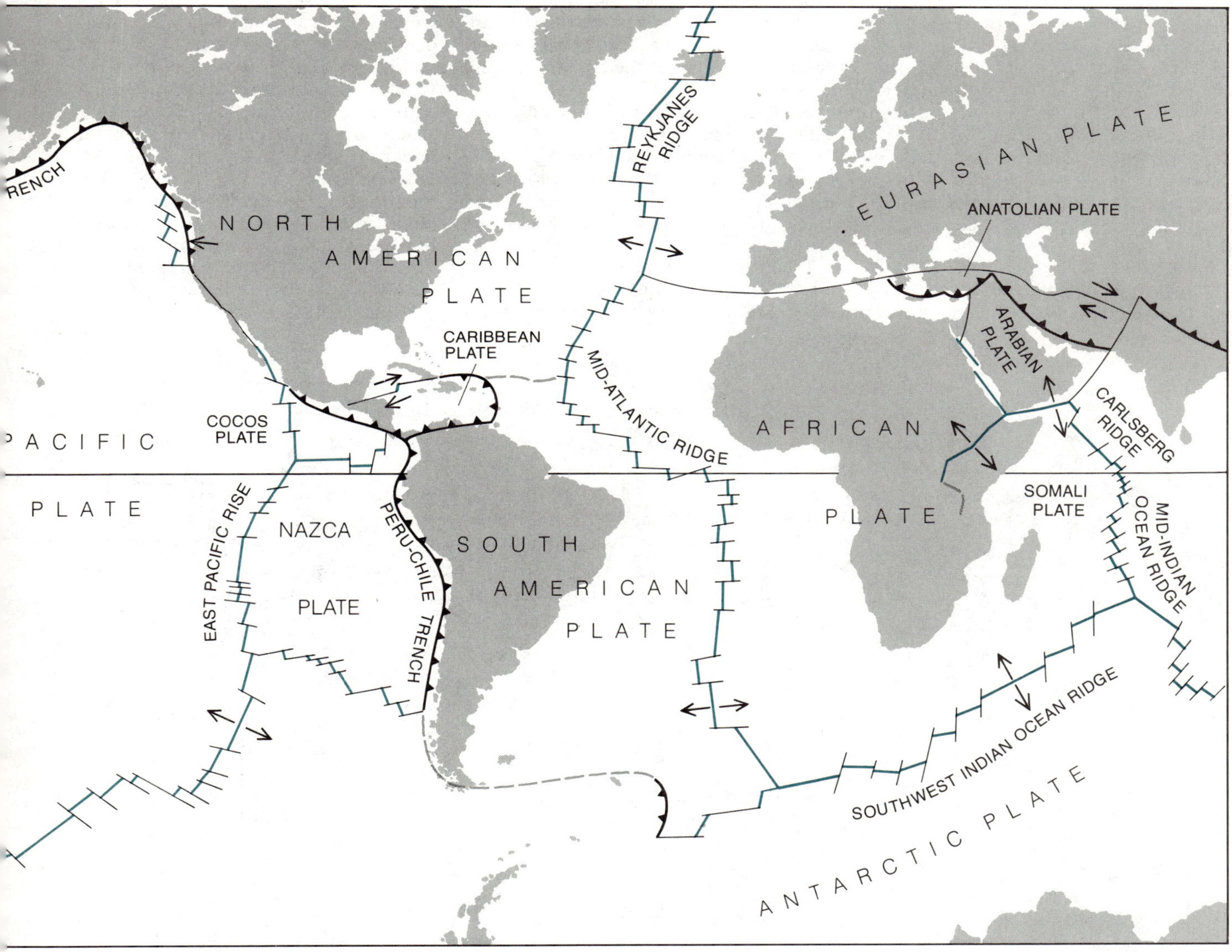

MAJOR LITHOSPHERIC PLATES and their boundaries are mapped. The lithosphere consists of the crust and the rigid upper part of the mantle. In midocean, away from the spreading center, the lithosphere is about 100 kilometers thick. The upper five to seven kilometers is crust. The lithospheric plates are in motion with respect to one another, and the boundaries are defined according to this relative motion. At divergent boundaries the plates move apart; the spreading centers at the midocean ridges are divergent boundaries. At transform faults the plates slide past each other. At convergent boundaries the plates move together and one plate plunges under the other in the process of subduction. Thus oceanic crust is created at divergent boundaries and destroyed at convergent boundaries.

tude of 800 kilometers, had surveyed the world's oceans between 72 degrees north and 72 degrees south.

The data from the three months of *Seasat* operation are currently being processed to yield dramatic images of the ocean floor [*see illustration on pages 42 and 43*]. The principle employed to convert the information on sea height into topographic maps is an intriguing one. The main cause of the spatial variation in the height of the sea surface is variation in the gravity field at sea level: the ocean tends to "pile up" where the gravity field is high and do the opposite where the gravity field is low.

The differences in gravitation are measured in relation to the ellipsoid. The ellipsoid is a mathematical figure corresponding to what the average sea surface would be if the mass of the earth were distributed in a radially symmetrical way.

The mass of the upper layers of the earth under the oceans is not, however, distributed in a radially symmetrical way. Hence the sea surface does not follow the ellipsoid. Instead it follows an irregular figure called the geoid. The difference between the geoid and the ellipsoid at any point on the ocean corresponds to the gravitational anomaly there. Where the geoid is higher than the ellipsoid there is a positive gravitational anomaly. Where the geoid is lower there is a negative anomaly.

What could account for the differences in the gravity field? Since the gravitational force exerted by a body is proportional to its mass, the presence of large topographic features such as seamounts, or undersea mountains, rising from the ocean floor is associated with positive gravitational anomalies; depressions or valleys in the floor are associated with negative anomalies.

In the *Seasat* maps of the ocean floor the fracture zones, volcanic seamounts and deep trenches near subduction zones stand out as if the ocean had been drained away. Their spectacular vistas aside, geotectonic images are being put to significant geophysical uses. For example, by providing detailed maps of fracture zones the geotectonic images are helping in the reconstruction of the direction of plate motion in the past.

The geotectonic images do not correspond exactly to the topography of the sea floor. Although seamounts, fracture zones and trenches are rendered with great clarity, one of the main spreading centers in the Pacific, the East Pacific Rise, appears as an insignificant swell. The reason for the discrepancy can be understood by considering again the block of wood. Imagine the block is a topographic feature on the crust and *Seasat* is flying over it comparing its gravity field to the fields around it.

When the block floats, part of it extends above the surface; hence there is an additional mass at that point on the surface and a positive gravitational anomaly would be expected there. The block is less dense than the fluid in which it floats, however, and so below the surface there is somewhat less mass than there is at the nearby points where no blocks are floating. When the block is at flotational equilibrium, the mass added above the surface and the mass subtracted below the surface are equal. Therefore under conditions of isostatic equilibrium in the crust a much smaller gravitational anomaly is observed than would be expected on the basis of the topography alone. The topographic feature is thus almost invisible to *Seasat*.

The midocean ridges are approximately in isostatic equilibrium, and so they appear as relatively small upswellings, much smaller than they are in reality. Other features are out of equilibrium and appear quite clearly. Consider a seamount on a piece of old oceanic crust. The crust has cooled and contracted and is quite rigid; hence the seamount cannot sink down far enough to reach equilibrium at that point. Instead the crust is depressed only slightly under the seamount but the shallow depression extends a long distance around the mount. Thus the large area that includes the seamount is in equilibrium but the point where the mount sits is not. The subtracted mass is spread over a considerable area of the crust, whereas the added mass is concentrated at the

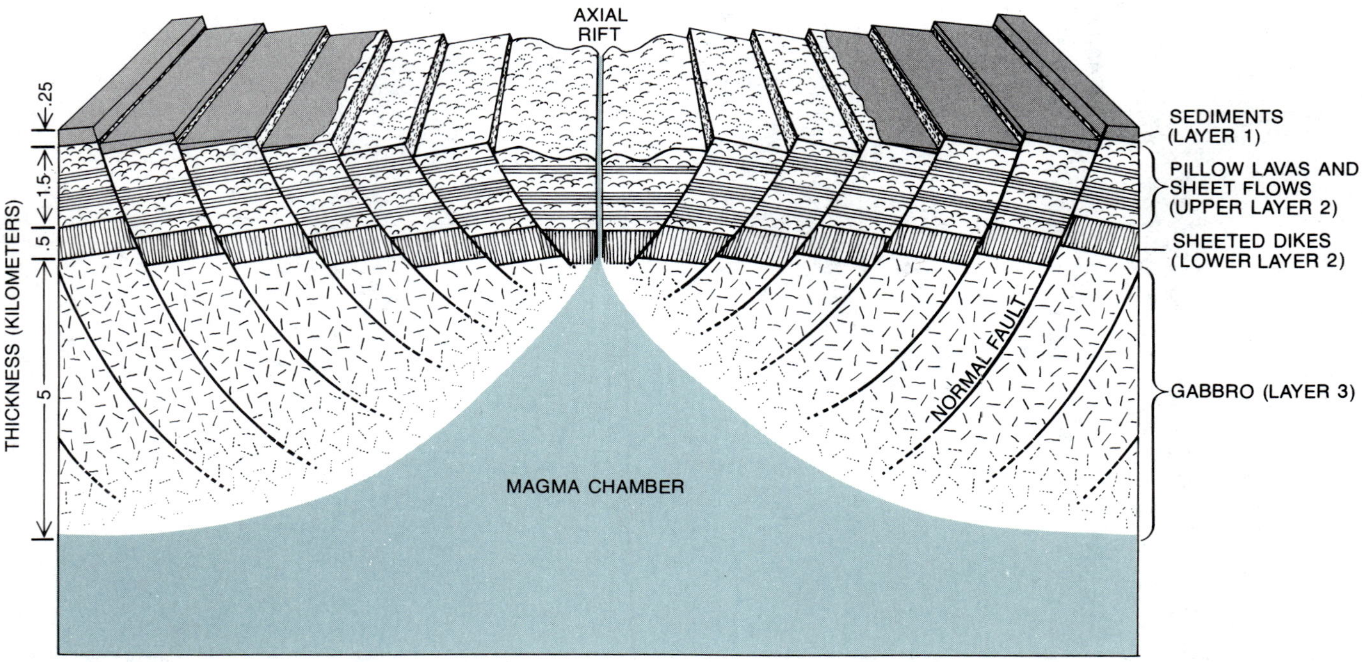

SPREADING CENTER is where magma is injected into the crust, as is shown in this cross section of the midocean ridge. The magma forms as the lithospheric plates diverge and mantle rock rises and melts with the decrease of pressure. It collects in a chamber below the spreading center. Within the magma chamber the rock gabbro crystallizes. At the top of the chamber the magma rises as the plates diverge and cools as vertical dikes. At the surface lava flows out and hardens in the form of sheets and "pillows." As the new crust moves away from the spreading center, a layer of sediments is deposited on it. The crust also cracks along normal faults, which run parallel to the ridge crest. Hence the mature crust has a layered structure from the top down: sediments, sheet and pillow lavas, dikes and gabbros.

seamount. The result is a positive gravitational anomaly at the seamount and a smaller negative anomaly around it.

Because the Mid-Atlantic Ridge and the East Pacific Rise are almost invisible to instruments that measure the gravity field, other techniques must be employed to map them. The most significant are methods based on recording acoustic pulses with a very high frequency projected from a device on shipboard to the ocean floor. The depth under the ship can be calculated from the travel time of the pulse from the ship to the bottom and back. In addition to the capacity to map spreading centers such sonic methods can provide much greater detail than a satellite can.

Sonic techniques are not new, but recent innovations have made them much more valuable to earth scientists than they were previously. The most significant innovation is arrays of sonar sources and receivers mounted in a single device. The information from the multiple beams can quickly be combined into a detailed topographic picture. An even more recent innovation is to mount such arrays on vehicles towed at considerable depth behind a ship.

The system called SeaBeam, made by the General Instrument Corporation, has an array of 16 beams, each with a frequency of 12,000 hertz (cycles per second). Each beam is a little less than three degrees wide. The beams are projected out in the shape of a fan along a line perpendicular to the ship's course, enabling the device to map a swath of sea floor two-thirds the water depth as the ship advances.

High-resolution SeaBeam mapping is a powerful tool for examining the sea floor, particularly the parts of the ridge system where the plates diverge rapidly. The spreading rate at the ridge crest can vary considerably. The Mid-Atlantic Ridge is a medium-to-slow spreading center with a rate of about five centimeters per year. The East Pacific Rise, on the other hand, is a fast spreading center where the rate of divergence can be as high as 15 centimeters per year. At fast spreading centers the features at the crest of the ridge tend to be concentrated in a fairly narrow zone that can be taken in with a single pass of SeaBeam; hence the system is particularly useful for mapping areas such as the crest of the East Pacific Rise.

The axis of the East Pacific Rise runs roughly north-south off the coast of Central America and South America. At its crest is a ridge about 30 kilometers across and 500 meters high. The axial ridge has recently been the subject of intensive work both by earth scientists interested in the fast spreading center and mining companies interested in the sulfide compounds deposited along the

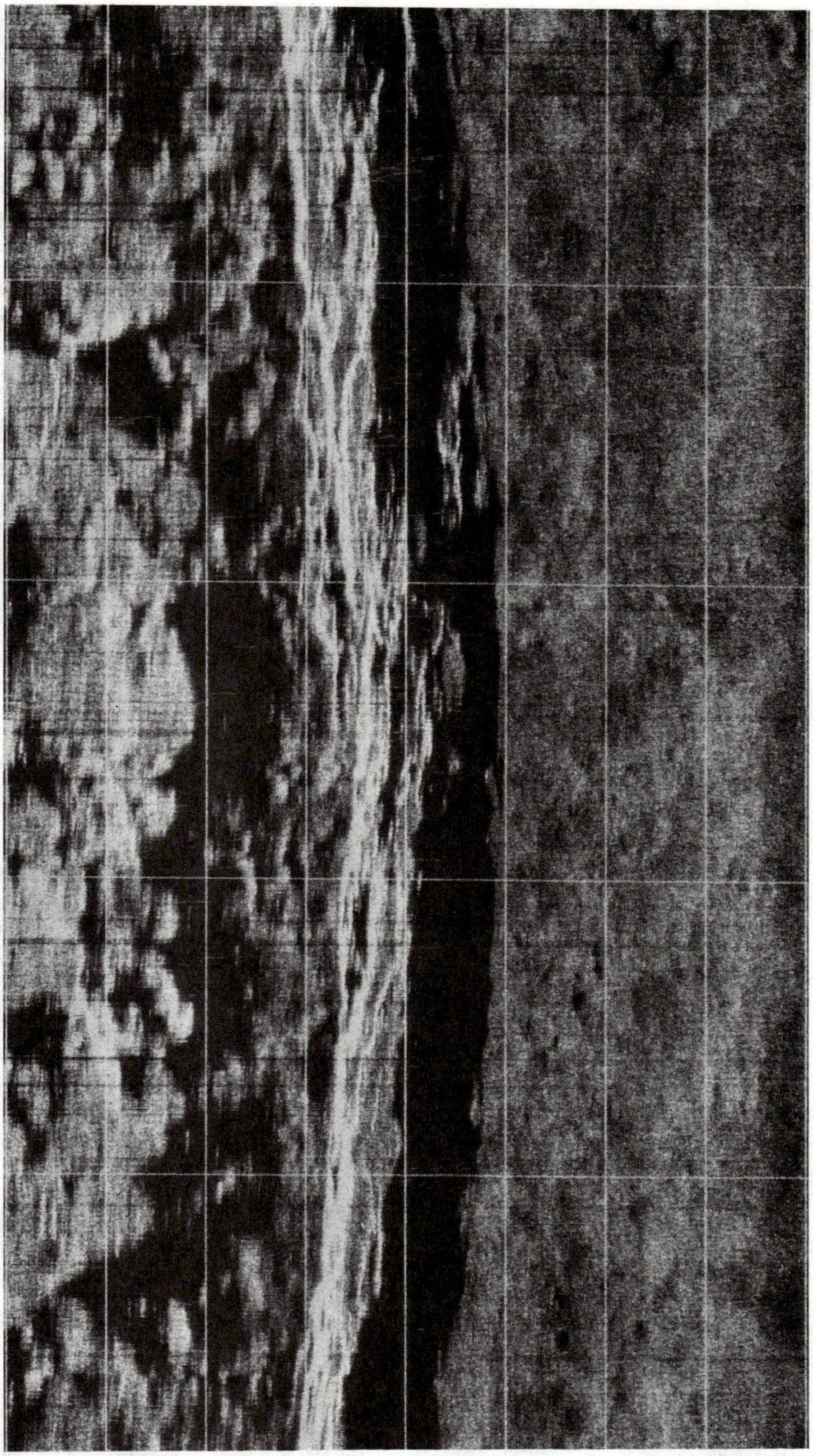

AXIAL RIFT delineates the exact center of the midocean ridge: it is the line along which lava flows out onto the surface. The rift shown is on the East Pacific Rise, a major spreading center that runs along the coast of Mexico, Central America and South America. The image was made with the sonar system SeaMarc I. A source of high-frequency acoustic pulses is housed in a "fish" that is towed close to the ocean floor behind a ship. The pulse is projected to the side and the intensity of the reflected beam is recorded. The data are later converted into an image of the ocean bottom. In this sonar image highly reflecting surfaces are light and nonreflecting surfaces and shadows are dark. The image was made during a cruise over the Clipperton Fracture Zone at 21 degrees north in the Pacific led by William B. F. Ryan of Lamont-Doherty.

BLACK SMOKERS are vents through which hot water is expelled from the crust. The particles that give the plume its dark color are sulfides that have been dissolved out of the crustal rock. Near the midocean ridge seawater penetrates cracks in the newly formed crust. The water is heated and propelled upward through the vents. The photographs on this page were taken on the East Pacific Rise from the manned submersible *Cyana*. This one was taken during an expedition led by Roger Hékinian of the Centre National pour l'Exploitation des Océans.

FISSURE CUTS THROUGH THE SURFACE of the oceanic crust near the crest of the East Pacific Rise. Such fissures are caused by plate motion, which generates lateral stresses, and also by the contraction of the crust. The fissures nearest the ridge crest are active sources of lava flow and are often concealed by the flows. Those farther from the ridge crest, like the one shown, no longer expel lava. This photograph was taken in 1978 during the first exploration of the East Pacific Rise by manned submersible in an expedition led by the author.

ridge by the heated seawater rising from the vents.

The crest of the ridge includes a zone of active volcanism. In the course of dives with the French submersible *Cyana* in early 1982 other investigators and I observed that the area of current volcanism is only one to two kilometers wide, so that it can easily be mapped in one SeaBeam pass. Indeed, about two years earlier a SeaBeam map of the "Great Pacific Highway," the narrow, flat crest of the rise, was made from the French research vessel *Jean Charcot*. Similar maps have since been made by other vessels, and the detailed topography of the East Pacific Rise is becoming clear.

The simultaneous employment of sonar mapping and manned submersibles on the East Pacific Rise has revealed a dramatic topographic pattern that could modify the notion of how the spreading center works. In the northern part of the East Pacific Rise major fracture zones marked by transform faults interrupt the ridge crest every 200 to 300 kilometers. Between the large faults are many smaller faults, some as little as 10 kilometers apart.

Near the transform fault the ridge is quite deep, and it rises to a peak between each pair of faults. The topographic high is usually equidistant from the two faults; the overall swell is about 500 meters over the 200-to-300-kilometer distance. Thus in a profile taken along the strike the ridge has the appearance of a gently rising hill. Since the profile of the ridge transverse to the strike is also that of a gentle rise, the area between a pair of transform faults is shaped like a low dome.

A second set of smaller domes protrudes from the large structure like a row of blisters along the ridge crest. This second set of domes was discovered in multibeam sonar studies by Robert D. Ballard of the Woods Hole Oceanographic Institution, Roger Hékinian of the Centre National pour l'Exploitation des Océans and me on the East Pacific Rise between the Orozco Fracture Zone at 15 degrees north and the Clipperton Fracture Zone at 10 degrees north. The information from the sonar maps was augmented by dives in *Cyana*.

The small domes are bounded by the faults that cut the ridge between transform faults. The small projections are about 100 meters high. It is probable that each dome corresponds to a spreading center: a small area where the manufacture of new crust goes on independent of the adjacent ridge segments. Thus instead of a single great factory the midocean ridge could be a string of small adjacent workshops. Hans Schouten and his colleagues at Woods Hole have suggested that the midocean ridge is indeed such a chain of adjacent spreading cells separated by fracture zones; the cells could be stable for long periods of time. Schouten's hypothesis implies that oceanic crust is not created as a single homogeneous mass but is made in long narrow ribbons laid side by side with fracture zones in between.

Although SeaBeam is the most convenient sonar system for mapping along-strike topography, two other sonar systems can give a more detailed picture of the ocean floor than SeaBeam can. Both are "side-looking" sonars, meaning that the acoustic pulses are projected to the side from devices towed behind the ship near the ocean floor. The Geological Long Range Inclined ASDIC (GLORIA) was developed at the Institute of Oceanographic Sciences at Wormley in England. The sonar transmitters encased in the "fish," a neutrally buoyant housing, emit pulses of sound with a frequency of 6,200 to 6,800 hertz. GLORIA can map a swath of sea floor about 30 kilometers wide.

A side-looking sonar system has one great advantage over a single-channel sonic profiling system operated from shipboard. In the shipboard systems vertical features such as scarps are difficult to detect because they yield little upward reflection of the acoustic beams. In the side-looking system, however, vertical features show up clearly because they offer excellent surfaces for horizontal reflection. Thus at the midocean ridges GLORIA has made plain the pattern of inward-facing scarps along the faults that parallel the ridge axis.

Each side-looking sonar system has its own frequency and arrangement of beams, hence it is best suited for a particular purpose. SeaMarc I was developed by International Submarine Technology, Ltd., to search for the remains of the liner *Titanic*. The SeaMarc device is towed in a fish from 100 to 400 meters above the sea floor. It transmits two roughly horizontal beams with a frequency of 27,000 to 30,000 hertz. SeaMarc can map a section of sea floor about five kilometers wide; its scale is thus between that of SeaBeam and that of GLORIA. Combining results from the SeaBeam, GLORIA and SeaMarc systems is beginning to provide a rich picture of the spreading centers. On a recent cruise led by William B. F. Ryan of Lamont-Doherty and J. Paul Fox of the University of Rhode Island the SeaMarc system was employed to map the East Pacific Rise between the Clipperton Fracture Zone and the Orozco Fracture Zone. The structures associated with the creation of oceanic crust could be seen as clearly as if they had been observed from the air.

In the past five years satellite data and sonar recordings have filled more detail in the map of the ocean floor than has been added in any comparable period before. As a result earth scientists now have almost as clear a picture of some parts of the midocean ridge as they do of structures on land. To penetrate below the crust and develop a picture of its composition and structure, however, other techniques are needed.

Much of the current understanding of the layers of the oceanic crust comes from recordings of seismic waves, both those generated naturally by earthquakes and those generated experimentally by explosions or special air guns. Indeed, the definition of the crust was originally formulated on the basis of work with seismic waves. The Mohorovičić discontinuity, or Moho, which divides the crust from the mantle, was first detected because of its capacity to reflect seismic waves.

The speed with which seismic waves travel depends on the temperature, pressure and composition of the rock medium. Hence recordings made at a distance from an earthquake or an experimental explosion can yield clues to the makeup of the intervening material. Two main types of waves are studied in the seismic work: body waves, which tend to travel through a particular layer, and surface waves, which tend to travel along the boundary between two layers.

Both body waves and surface waves can consist of two other types of waves: *P* waves and *S* waves. *P* waves are analogous to sound waves in air. As the *P* wave passes, the rock is compressed and expanded only in the direction of wave motion. *S* waves subject the rock to shear stresses perpendicular to the direction of wave motion. *S* waves can travel only through solids; *P* waves can travel through solids, liquids and gases.

In the late 1950's Russell W. Raitt of the Scripps Institution of Oceanography proposed a seismic model of the crust. In Raitt's model the crust was composed of three layers that could be distinguished on the basis of the *P*-wave velocity. Layer 1 consisted of the sedimentary cover, Layer 2 the oceanic basement and Layer 3 the oceanic layer. The underlying mantle was designated Layer 4. The *P*-wave velocity was held to be constant in each layer.

Raitt's hypothesis included three simplifying assumptions. The first was that the wave velocity in the rock always increases with depth. In general this is the case, because the compression of the rock increases its "resonance" as a medium for wave motion. There are, however, exceptions to the rule, and the velocity pattern in the crust is complex.

Raitt's second assumption was that the boundary between any two layers of the crust is a horizontal plane. His third assumption was that the layers are quite thick in relation to the characteristic length of the waves. In general a seismic wave can give information only about features that are considerably larger than the length of the wave. Features that are smaller than the wavelength have little effect on the path of the wave and therefore cannot be detected when the wave motion is analyzed. Experimental explosions (which were the main source of seismic waves in geologic work in the 1950's) generally give rise to waves with a wavelength of from .5 kilometer to two kilometers; thus they yield information only about components of the crust that are a few kilometers thick or thicker.

Recent work has shown that all three assumptions are drastic oversimplifications. Even the idea that the velocity in the layers is constant has now been called into question. Paul Spudich and John A. Orcutt of Scripps and G. Michael Purdy of Woods Hole have put forth a model of a crust in which layers are defined not by the absolute wave velocity but by the velocity gradient: the change in wave velocity with depth. Recent seismic work has been much more concerned with Layer 2 and Layer 3 than with Layer 1, where sediment often gives inconsistent seismic results.

According to Spudich and Orcutt, in Layer 2 the velocity gradient is quite steep. The wave velocity increases by from one to two kilometers per second with each kilometer of depth in the crust. In mathematical notation such a gradient is written as 1 to 2 s^{-1}. In Layer 3 the gradient is about .1 s^{-1}. Thus the *P*-wave velocity in Layer 3 is almost uniform, which makes Layer 3 the best defined region of the crust from the seismic point of view. In Layer 4, the upper mantle, the wave velocity exceeds 7.8 kilometers per second.

Some seismologists argue that there is an additional low-velocity zone between Layer 3 and Layer 4 corresponding to the transition from the crust to the mantle, but the evidence for such a layer is scanty. It is known that the transition from the crust to the mantle takes place over a distance of three or four kilometers. Waves with a wavelength short enough to give a high-resolution picture of the Moho, however, become quite attenuated in traveling to the bottom of the crust and back to the surface, and therefore little is known in detail about the transition layer.

Purdy took a seismic profile southwest of Bermuda in oceanic crust created during the Mesozoic era about 140 million years ago. Carefully allowing for the effects of a variable sediment cover, he found that the crust was 7.2 kilometers thick above a 500-meter transition zone to the mantle. The *P*-wave velocity at the top of the crust was five kilometers per second. Layer 2, defined as the region where the gradient was greater than .64 s^{-1}, was 2.3 kilometers thick. Layer 3, with a gradient of .1 s^{-1} or less, was 4.9 kilometers thick.

In the crust where Purdy did his work Layer 3 is subdivided into two levels: an upper level 1.7 kilometers thick with a gradient of .1 s^{-1} and a lower level 3.2 kilometers thick where the waves show no change in velocity with depth. The velocity in the lower level of Layer 3 is seven kilometers per second. Thus the oceanic layer (Layer 3), seismically defined, is about five kilometers thick and the basement layer above it (Layer 2) is half as thick.

Several workers have attempted to go beyond the structural profile given by the velocity gradients to infer the composition of the rock layers. This is generally done by comparing the *P*-wave velocity with the *S*-wave velocity for rock at a given depth in the crust. Different kinds of rock have characteristic ratios of *P*- to *S*-wave velocities and hence some deductions about the rock type can be made on the basis of the seismic data. Unfortunately for the seismologist the wave-velocity ratios for the rock types are not unique. The velocity change from Layer 2 to Layer 3 could therefore be interpreted as either a change in metamorphic rock type (from greenschist above to amphibolite facies below) or a change in lithologic type (from basalt above to gabbro below).

It is possible that knowledge of the exact composition of the lower layers of the oceanic crust will have to wait until drills penetrate deeper in the crust than they have done so far. In the winter of 1981 a drill on the *Glomar Challenger* in the six-nation International Program for Ocean Drilling succeeded for the first time in going deeper than one kilometer. The drill hole, 1,076 meters deep and designated 504-*B*, is in the Costa Rica rift, which crosses the East Pacific Rise between the Galápagos Islands and South America.

Although a one-kilometer hole seems modest compared with the full seven-kilometer depth of the crust, the achievement is a considerable one because of the problems of drilling in the ocean. The difficulties encountered in drilling holes in the oceanic crust have included the twisting and breaking of the drill string and severe wear on the bits. The drilling of Hole 504-*B* was relatively free of such problems. What does Hole 504-*B* tell us about the crust?

The crust at the drill hole is six million years old. Because the Pacific at the latitude of the Costa Rica rift is warm and biologically active, the surface of the crust is already covered with a layer of sediment 275 meters thick. Below the sediments the upper 575 meters of the basement consists of pillow lavas and the rocks called breccias and hyaloclastites, which are produced by the melding of small pieces of fractured basalt into a single mass under pressure. Between 575 and 780 meters the first dikes appear and extensive breccias are interspersed with a few pillow lavas. From 780 meters to the bottom of the hole are massive basalts, abundant dikes and a notable lack of either pillow lavas or fractured material.

Recordings of the in situ pressure deep down in Hole 504-*B* made by Roger N. Anderson of Lamont-Doherty and Mark Zoback of the U.S. Geological Survey are clarifying the circulation of water through the upper layers of the crust. Anderson and Zoback employed an inflatable device called a packer to isolate a section of the hole. The pressure in the isolated zone was measured from a "go devil" developed by the oil industry to explore oil and gas wells. Intriguingly, Anderson and Zoback observed that the water halfway down the hole is under less pressure than the water at the top. The difference in pressure is about eight bars. (One bar is 14.7 pounds per square inch.)

The mechanism of the remarkable reversal of the pressure gradient is not yet understood. The underpressure in the hole could, however, be due to a heat-convection cycle in the mantle. The relatively low pressure in the lower layers of the crust could help to explain the hydrothermal circulation; it would tend to pull seawater from the ocean floor down into the cracks in the crust.

The depth of the hydrothermal circulation is currently a matter of considerable controversy. It has been suggested that the circulating water penetrates deeply enough to have a role in regulating the operation of the magma chamber under the midocean-ridge axis. If enough seawater reaches the lower layers of the crust, it could cool the magma and cause it to solidify. The lava would therefore stop flowing onto the surface of the ridge until the plates pulled apart enough to renew the decompression of the mantle rock.

For the circulating water to have such an effect it would have to penetrate into the gabbro in Layer 3. Hole 504-*B* did not go deep enough to settle the question, but it is significant that the rocks recovered from a depth of 600 meters on down to the bottom of the hole show a pattern of alteration that is quite compatible with the fluxing of heated seawater through them. Moreover, the rocks from the deepest part of the hole are among the most extensively altered. Such alteration would have had to take

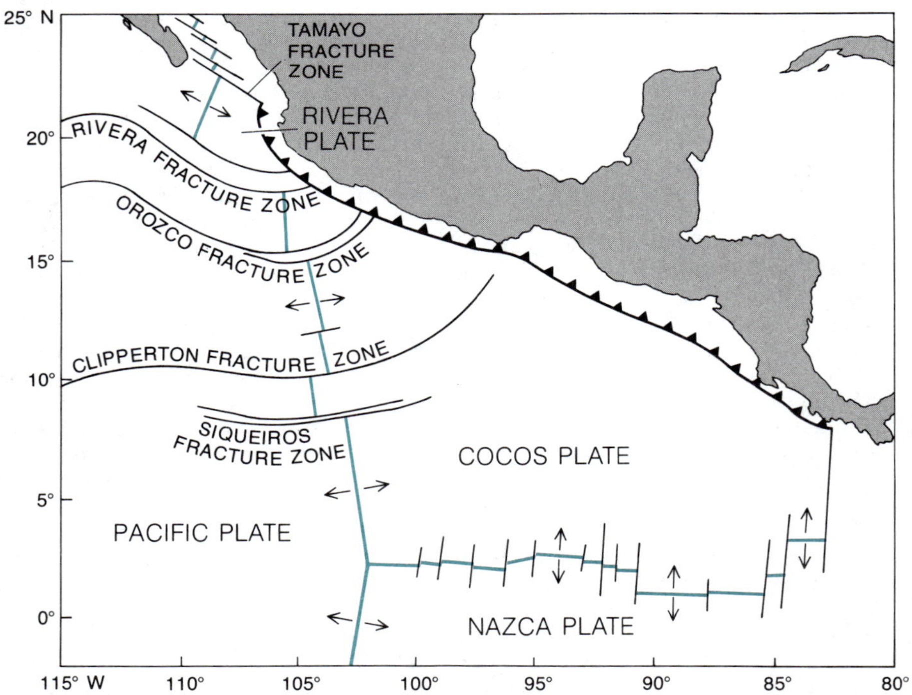

EAST PACIFIC RISE OFF SOUTHERN MEXICO has a complex geometry that includes many faults across the strike, or main axis, of the ridge. The axis of the rise runs along the boundary between the Pacific plate and the Cocos plate; arrows indicate the direction of plate motion. Fracture zones mark the points where major transform faults cross the strike. At the fracture zone the ridge axis is offset. Within the offset the direction of movement of the crust on one side of the fault is opposite to that of the crust on the other side. Between the fracture zones are many smaller faults. Much of the author's recent work in the *Cyana* and the research vessel *Jean Charcot* has been done between the Clipperton and Orozco fracture zones.

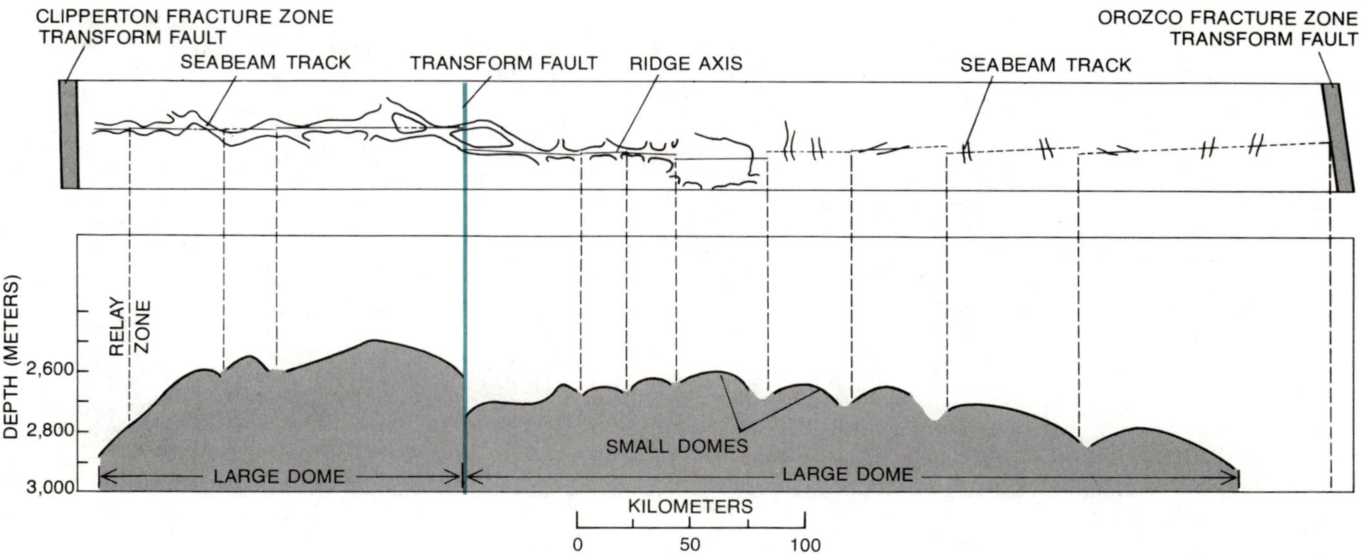

SUPERPOSED DOMES at the crest of the East Pacific Rise could give clues to how the spreading center at the midocean ridge operates. The upper panel shows part of the East Pacific Rise in plan view as mapped by the SeaBeam sonar system. The mapping was done by the author with Hékinian and Robert D. Ballard of the Woods Hole Oceanographic Institution. The ridge axis is marked by the central line; broken segments stand for regions where the axis had to be guessed at. The axis is marked by active volcanic fissures. Pairs of parallel lines show areas that have been mapped by the SeaBeam sonar system. Ridge offsets mark transform faults and relay zones, where small faults cross the strike. The lower panel shows the same ridge segment in profile. Two large domes run between transform faults, with the peak about halfway between the faults. A series of small domes bounded by relay zones protrude from each large dome.

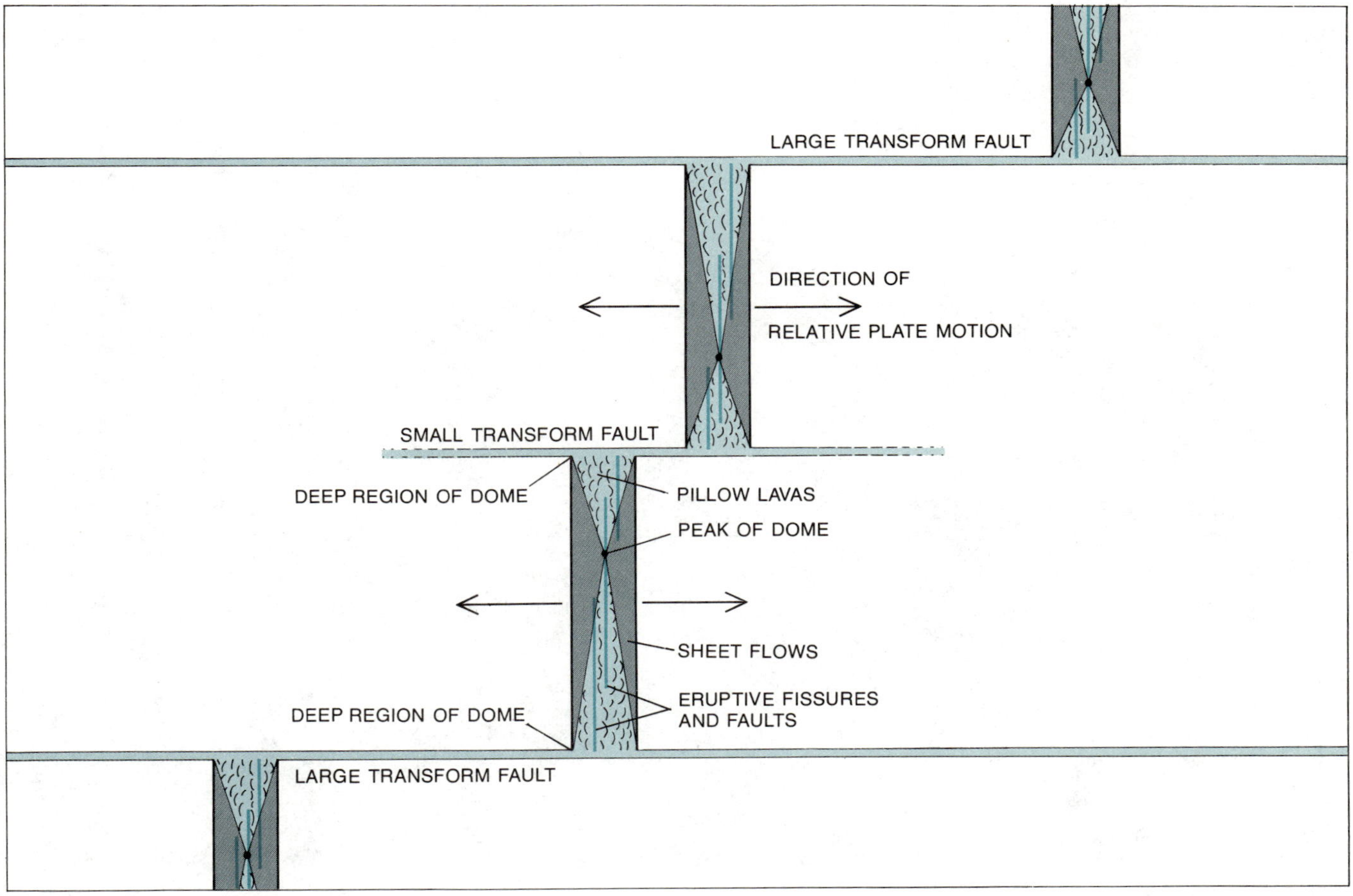

HYPOTHESIS proposed by the author and Ballard to explain how the spreading center operates includes the idea that each large dome between transform faults is an individual spreading cell: an area where the creation of oceanic crust proceeds independently. In this schematic diagram two complete domes are shown. The broken line of eruptive fissures from which lava flows marks the ridge crest. Each dome has a peak between faults; the dome slopes down to deep regions next to the fault. Near the peak the ridge crest is covered mainly by sheet flows. Farther down, the slope is covered by pillow lavas. Which form of lava is predominant depends on the position on the dome. If each dome is a spreading cell, the crust could be created in thin ribbons extending from the sides of the spreading cell.

place before the sediments were deposited, capping the basement and preventing the flow of water into it.

The clues to the operation of the magma chamber gained from Hole 504-*B* are being supplemented by seismic work. In investigating the magma chamber both the reflection and refraction of seismic waves are observed. In the reflection experiments air guns towed behind a ship are generally employed as the energy source. The waves from the air guns pass down to the crust and are reflected upward to the ship, where their travel time and amplitude are recorded. In the refraction experiments the energy propagates at the interface between rock layers and can be recorded some distance away on the sea floor with an ocean-bottom seismometer or at the sea surface. By combining reflection and refraction results the speed of the waves through the structures of the crust can be calculated.

The speed of seismic waves is reduced considerably by rock in the molten state; therefore a low-velocity zone in the crust could correspond to a magma chamber. There is an area under the East Pacific Rise where the refracted waves are attenuated or much slowed; the area is probably a crustal magma chamber. Magma is an efficient reflector of seismic energy, so that in the reflection work the top of the magma chamber comes out as a strong, roughly flat reflecting surface two or three kilometers below the sea floor. In the spreading center on the East Pacific Rise at nine degrees north and also in the Lau Basin near the Fiji Islands the reflection at the top of the chamber is about four kilometers wide.

Refraction experiments done by Brian T. R. Lewis and his colleagues at the University of Washington also make it clear that the magma chamber is quite narrow. Lewis believes it is even less than four kilometers wide. Thus the entire oceanic crust is created by a thin tube of molten rock running a few kilometers below the crest of the midocean ridge.

The "root" of the midocean ridge, the underlying structures in the mantle, is also being seismically probed. One means of examining the deeper layers is the surface waves called Rayleigh waves. Donald W. Forsyth of Brown University, Nicole Girardin of the Institut de Physique du Globe of the University of Paris and Wolfgang Jacoby of the University of Frankfurt studied the passage of Rayleigh waves through young lithosphere in the Pacific and also along the Reykjanes Ridge south of Iceland. They found that the low *S*-wave velocity extends down to 60 kilometers.

Additional data were supplied by waves from a strong earthquake in Uzbekistan in the southern U.S.S.R. in May, 1976. Rayleigh waves from the earthquake with a period of 300 to 400 seconds, which can make several circuits of the earth, were recorded with ultralong-wavelength seismometers at Los Angeles and at the South Pole. The passage of the long-wavelength waves shows that the low-velocity zone under the East Pacific Rise extends to a depth of 200 kilometers, with the minimum velocity at about 100 kilometers. Thus the midocean ridge has deep roots.

Stuart A. Hall of the University of Houston and his colleagues have advanced a quantitative model of the magma chamber and its root in the mantle. Their work is intended to account for the small gravitational anomalies on the crest of the midocean ridges. As I mentioned above, the ridge is approximately in isostatic equilibrium and so is not associated with large anomalies in the gravitational field. There are, however, small anomalies directly over the ridge crest. At the Mid-Atlantic Ridge, where the spreading rate is low, there is a negative anomaly; at the East Pacific Rise, where the spreading rate is high, there is a small positive anomaly.

According to Hall and his colleagues,

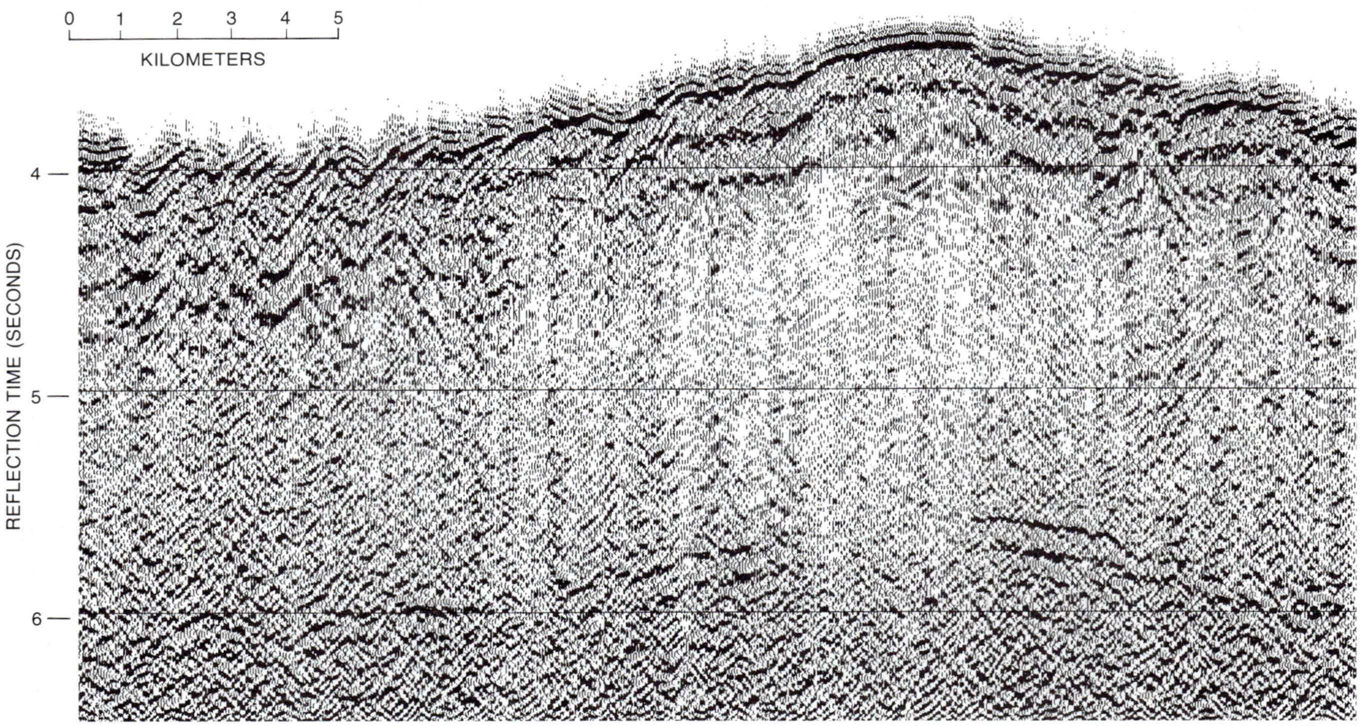

SEISMIC PROFILE offers evidence that the magma chamber under the midocean ridge is quite narrow. The profile was made by Peter Buhl of Lamont-Doherty at nine degrees 30 minutes north on the East Pacific Rise. The peak on the upper border of the profile is the ridge crest. The dark line at a reflection time of about six seconds is the Mohorovičić discontinuity, or Moho: the boundary between the crust and the mantle. A seismic profile is made by generating strong acoustic waves and recording their reflections from various levels of the crust. The travel times and amplitudes of the reflected seismic waves yield information about the character of the crustal rock. In a partially liquid magma the seismic waves are considerably retarded. The break in the Moho directly under the ridge crest indicates the presence of magma. The width of the break suggests that at the level of the Moho the magma chamber is less than two kilometers across.

the gravitational anomalies could both be explained by one type of magma chamber. They hypothesize that the density of the material in the magma chamber is about 2.75 grams per cubic centimeter, or 1 percent lower than that of the surrounding rock. The mantle root that is made up of gabbro has a density of three grams per cubic centimeter, or 6 percent lower than that of the surrounding mantle rock. The density of the chamber and the mantle root is thus quite close to the density of the adjacent rock.

Hall and his co-workers conclude that the small gravitational anomalies over the ridges are therefore due not to variations in density but to topographic features at the ridge crest. On the slow-spreading Mid-Atlantic Ridge there is a rift valley along the ridge axis. On the fast-spreading East Pacific Rise the axis is marked by a ridge.

The newest source of information about the structure of the oceanic crust is not seismic waves or gravity data but electromagnetic radiation. It is now possible to measure undersea electric and magnetic fields with considerable accuracy by means of receivers on the ocean bottom. If a source of electric current is put on the sea floor some distance away from such a receiver, the electric field induced by the current travels through the rock. The measured intensity of the electromagnetic indicates how well the intervening section of the crust conducts electricity.

The electrical conductivity of rock is affected by chemical composition, temperature and the extent of melting. A deep electrical conductivity log can therefore be quite informative about the rock of the crust and the upper mantle. As a source of electromagnetic energy Charles S. Cox and Peter Young of Scripps have employed an insulated wire 800 meters long with bared ends. The wire acts as a horizontal electric dipole, with the return flow of current being through the ocean. The wire is laid on the sea floor at the end of a cable attached to a ship. An alternating current with a peak of about 70 amperes is passed through the wire. The energy transmitted by the dipole has a frequency of about one hertz. A pair of cruciform antennas with arms nine meters long are put on the sea floor 19 kilometers away from the transmitter. With this setup Cox and Young have recorded electric field signals as strong as 10^{-10} volt per meter. Since the "noise," or background electric field on the sea floor, is 10^{-12} volt per meter, 100 times weaker than the recorded signal, the finding is significant.

The pattern of received signals shows that there are two layers in the crust with quite different electrical conductivity. The upper layer is at most 1.5 kilometers thick and has a fairly high electrical conductivity: about .1 Siemens per meter. The upper layer corresponds to the relatively young fractured basalts found near the top of Hole 504-*B;* the measured conductivity in the two places is about equal. In both places the presence of seawater that has penetrated the fractures greatly increases the electrical conductivity.

Below the conductive layer is a region, extending down to about six or seven kilometers, in which the conductivity is much lower: about .004 Siemens per meter. The measured conductivity in the lower region is probably an average for the lower crust and upper mantle. In the deep parts of the crust the conductivity is due to seawater penetrating the gabbro layer. In the mantle the conductivity results from the passage of the electric signal through minerals in the hot rock.

Electromagnetic measurements can yield unique information about the changes in temperature with depth and the presence of molten rock in the deep regions of the crust. Therefore this new method could in the next few years throw much light on the zone near the Moho, which is not easily investigated with other techniques. Novel and intriguing as it is, electromagnetic observation is only one of an array of methods that are currently being harnessed to probe the upper part of the earth under the ocean. Theories about the oceanic crust advance, even more than theories in other areas of science, only in conjunction with new methods of observation. The past decade has been a period of remarkably rapid development of such methods. Given the inevitable lag between observation and the formulation of new theories, in the next decade there could be developed a new and more accurate picture of the thin layer that covers most of the surface of the earth.

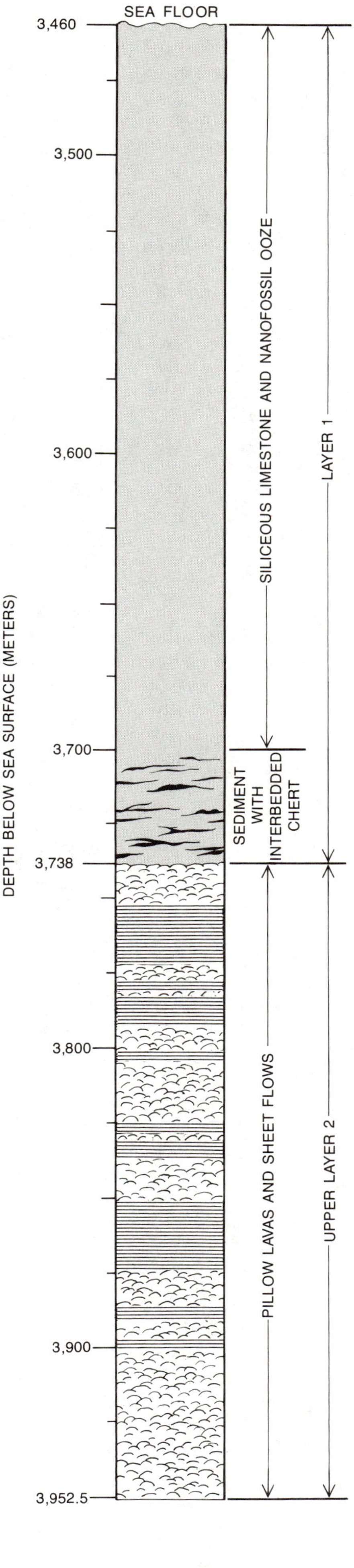

DEEPEST HOLE yet drilled in the oceanic crust is Hole 504-*B* on the Costa Rica rift between the Galápagos Islands and South America. Drill cores from Hole 504-*B*, such as the one shown in schematic form, are providing new information about the structure of the crust. The depth below the sea surface is given at the left; the core depicted extends for about half the full depth of the hole. The crust on the Costa Rica rift is about six million years old. It is covered by a sediment layer 275 meters thick, made up mainly of the remains of microscopic marine plants and animals. Below 275 meters is the oceanic basement, composed of pillow lavas and sheet flows. Intriguingly, it has been shown that the in situ pressure at the bottom of the core in the illustration is less than the pressure at the ocean floor. The difference in pressure could pull seawater down into cracks in the crust.

5

THE CONTINENTAL CRUST

The Continental Crust

by B. CLARK BURCHFIEL

It is much older than oceanic crust, some of it dating back nearly four billion years. It is constantly reworked, however, by cycles of tectonics, volcanism, erosion and sedimentation

For a little more than 200 years earth scientists have been studying the geology of the continents, hoping to reconstruct a record of the history of the earth. It is a daunting project. The crust underlying the oceans is rapidly created; it remains intact and relatively undeformed for most of its short lifetime, and then it is rapidly destroyed. The oldest crust in the earth's ocean basins today is less than 200 million years old. In contrast, the crust making up the continents is created and modified by a variety of physical and chemical processes, often undergoing many phases of deformation and reworking that produce a complex worldwide pattern in which belts of deformed rock hundreds of kilometers wide and thousands of kilometers long are invaded by intrusions of igneous rock and locally overlain by a thin veneer of younger sedimentary rock. In addition continental crust largely resists the processes that destroy oceanic crust. The most ancient parts of the continents are about 3.8 billion years old. Thus the continental crust holds a complex and fragmentary record of the evolutionary and dynamic processes operating through 85 percent of the earth's 4.6-billion-year history.

Continental crust underlies the continents and their margins, and also small shallow regions in the oceans. In total it covers about 45 percent of the earth's surface and makes up about .3 percent of the mass. It is distinguished from oceanic crust and from the mantle under it by its physical properties and chemical composition. The horizontal boundaries between continental crust and oceanic crust are poorly defined; they are under not only the ocean's water column but also, in most places, a thick sequence of sedimentary rock. Seismic, magnetic and gravitational data indicate that the boundary is less than 10 kilometers across in some places but is several tens of kilometers across in others. Studies of the rock of oceanic and continental crust and the correlation of rock compositions and seismic velocities indicate, however, that the oceanic crust is characterized by the igneous rock basalt whereas the continental crust is an assemblage of igneous, metamorphic and sedimentary rocks enriched in elements such as potassium, uranium, thorium and silicon.

The vertical boundary between the mantle and the crust (both oceanic and continental) is called the Mohorovičić discontinuity, more commonly the Moho. It is a zone, less than one kilometer thick in some places but several kilometers thick in others, where the velocity of compressional seismic waves increases from about 6.8 kilometers per second in the crust to 8.1 in the mantle. The change in seismic velocity is caused largely by a change in the composition of the medium. Rocks of the mantle differ from rocks of the crust: they are poorer in silicon but richer in iron and magnesium.

Seismic studies of the Mohorovičić discontinuity indicate that the oceanic crust is typically from five to eight kilometers thick, whereas the continental crust varies from about 10 kilometers to more than 70. To a first approximation the crust behaves as if it were floating on the mantle. Oceanic crust is relatively thin and dense (3.0 to 3.1 grams per cubic centimeter); thus the parts of the earth's surface underlain by oceanic crust are usually far below sea level, at a depth ranging from 2,500 to 6,500 meters. Continental crust is thicker and is notably less dense (2.7 to 2.8 grams per cubic centimeter); thus the parts of the surface formed by continental crust lie near sea level or above it. The thickest parts of the continental crust usually underlie places of great elevation, such as the Himalayan and Andean mountain chains. Conversely, the thinnest parts of the continental crust usually lie below sea level at places such as the continental margins of the Atlantic.

Some important exceptions to this pattern are found at midocean ridges and some areas on the continents where volcanism is active and the crust is being stretched. In such places hot material from the deeper mantle rises to shallower levels, making the upper mantle hotter and so less dense than normal. The resulting buoyancy raises the surface elevation. The Basin and Range province of the western U.S. is a good example. The crust there is thin, but the surface elevation is nonetheless high.

Other exceptions to the pattern are found in areas with great topographic relief, where the crust bends downward over short horizontal distances, usually about 200 kilometers. Evidently the crust and the uppermost mantle deform like an elastic sheet to support the topographic load. One result is that long, linear troughs filled with sediment as much as eight kilometers deep form next to many great mountain chains. The troughs are underlain by crust of normal thickness. The crust has simply bent downward to support the weight of the mountains.

In general the rocks that form the continental crust fall into two major groups: widespread, relatively undeformed accumulations of sedimentary or volcanic rocks on the one hand and long, deformed belts of sedimentary, igneous and metamorphic rocks on the other. The belts are called orogenic belts, after the Greek *oros*, mountain. The first

LABRADOR FOLD BELT, shown in a Landsat image made above north Quebec, exemplifies the evolution of the continental crust: the belt was once a chain of mountains raised by the collision of two continents 1.8 billion years ago. Since then the mountains have eroded, exposing the deeper, mostly metamorphic and igneous rocks that the collision had deformed into myriad folds. Moreover, subsequent plate collisions have reshaped the continents. Nevertheless, the belt resembles the ones being formed by tectonic activity along continental margins today.

group is not ubiquitous on the continents, but where it is present it always overlies the second. In places such as the central U.S. it forms a thin veneer no more than a few kilometers thick. Elsewhere, along continental margins and in linear, circular or irregular depressions in the continents, it forms sequences of rocks that can be more than 10 kilometers thick.

The second group makes up the bulk of the continental crust. Its great variety of rock assemblages gives it a heterogeneity contrasting significantly with the relative homogeneity of the oceanic crust. Each belt in the second group evolved over a span of time as long as several hundred million years, and the ages of adjacent belts may differ by hundreds of millions of years, or even a billion years; hence each belt represents a different segment of earth history. Often the younger belts are oblique to the older ones, so that the younger truncate the older. In other places belts are parallel.

A detailed examination of the rocks in the belts shows that many older belts are similar to the ones formed in the more recent geologic past. They also

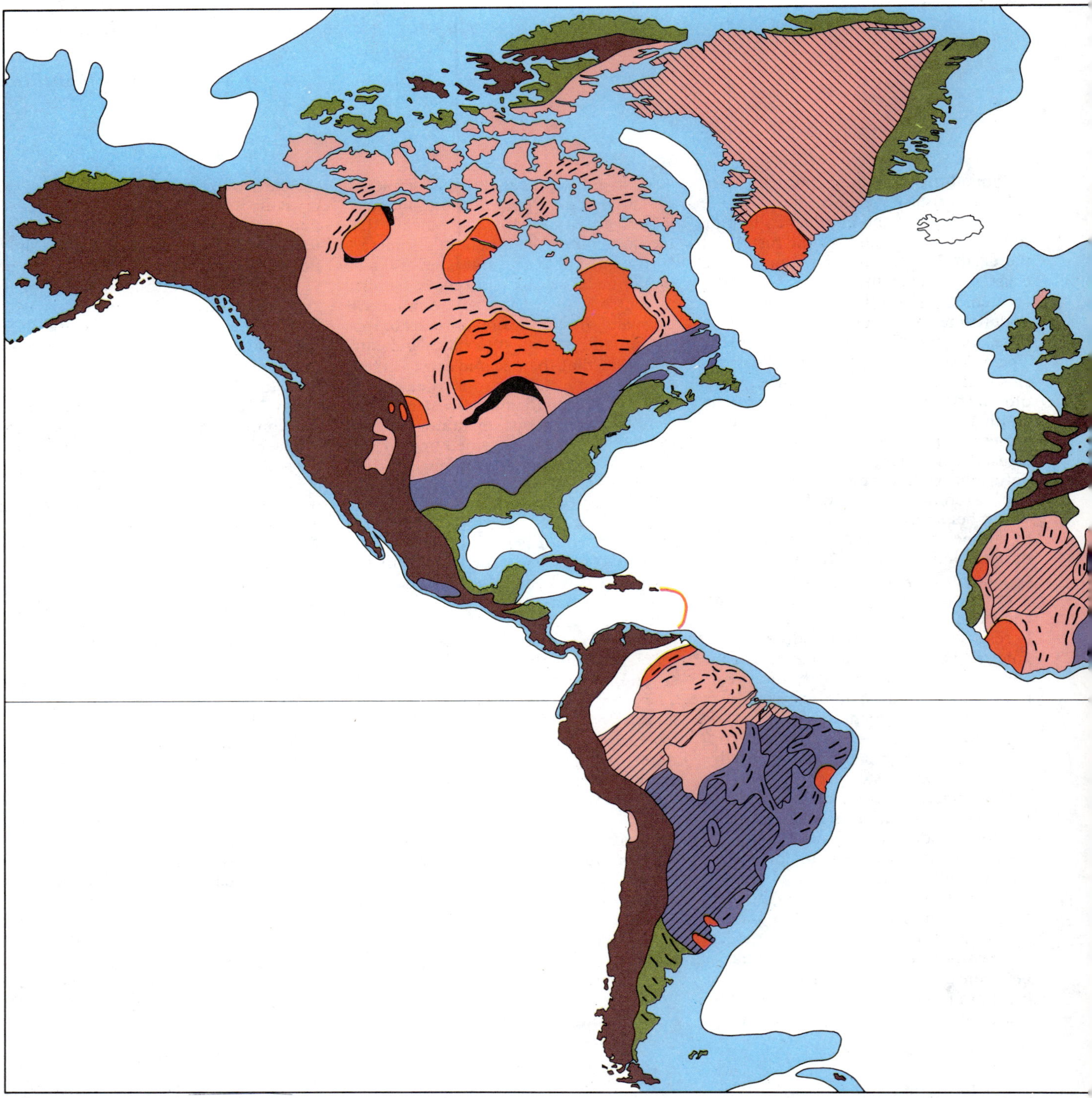

WORLDWIDE PATTERN OF BELTS of rocks deformed during different geologic periods accounts for the major proportion of the continental crust. In places the belts are overlain by sedimentary or volcanic rock or by ice (*black hatching*). Even then drilling and other methods of exploration confirm that the belts are present at depth. Each belt is complex: it records many individual episodes of deformation and usually includes rocks reworked from older belts. Col-

resemble the belts that the earth's tectonic activity is forming today. This offers the prospect that the study of modern rock formations and how they are currently deforming will yield understanding of the processes responsible for the formation of the older orogenic belts making up most of the continental crust. The theory of plate tectonics is crucial in such a venture, because the theory provides a framework in which rock assemblages and their deformation can be related to interactions of the plates that make up the entire crust. To be sure, the theory was developed largely from data gathered in the oceans, and its application to the study of continental crust has met with mixed results. Still, a modified version of the plate-tectonic concepts is a basis for understanding continental development.

The central concept of plate tectonics is straightforward: the outermost shell of the earth, the lithosphere, can be divided into six major plates and many smaller ones that move with respect to one another with velocities ranging from a few centimeters per year up to 20 or more. The plates consist of oceanic and continental crust together with

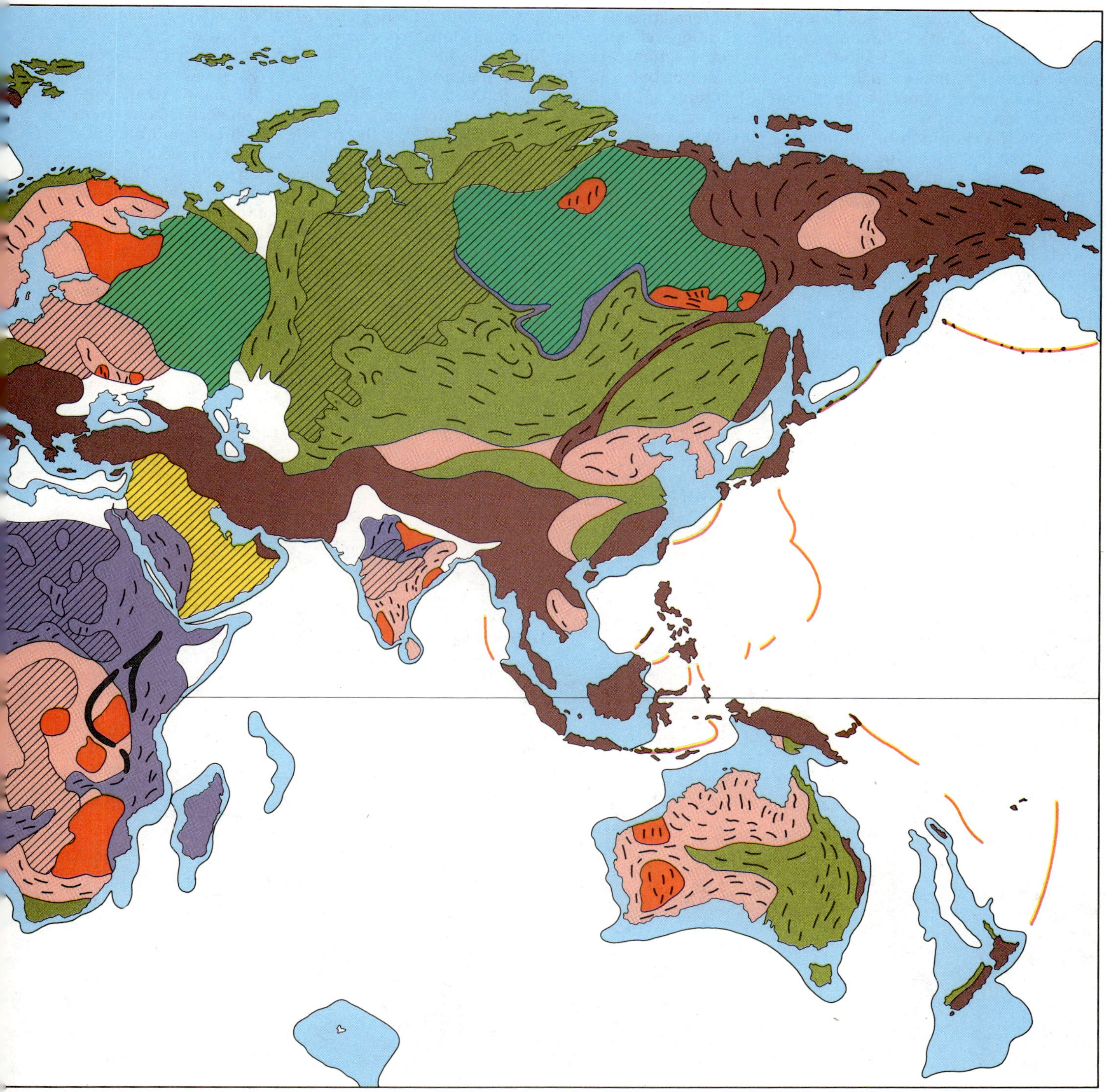

ors show the time spans of major episodes of deformation. Rocks of similar continental composition underlie the continental margins and parts of small oceanic plateaus and rises (*light blue*). In addition volcanic arcs (*orange*) built up on the oceanic crust form a type of crust that can be sutured to a continent when two plates slide past each other or collide. Rocks and structures arising from the divergence of two plates are also preserved in the continents. Active rifts such as the East African rifts, older rifts such as the Oslo rift in Scandinavia and ancient rifts such as the Athapuscow and Bathurst rifts in Canada, which are more than two billion years old, are examples (*gray*).

some of the underlying mantle; the Mohorovičić discontinuity lies within them. The plates are generally regarded as rigid bodies, so that most of their interactions are concentrated along plate boundaries, which can be zones of intense deformation. The boundaries can be classified into three basic types: divergent, transform and convergent. At divergent boundaries new oceanic crust is created; at transform boundaries it slides past the crust of a neighboring plate; at convergent boundaries it plunges into the mantle. The continental crust generally resists this subduction, largely because it "floats" on the mantle.

This basic scheme must be modified in several important ways if it is to help illuminate the evolution of continental crust. In the first place the continental crust participating in plate interactions often does not behave rigidly. In the upper lithosphere, therefore, the motions of plates may be absorbed partially or entirely by deformation within the continental crust. Much of the deformation can be attributed to forces arising at plate boundaries; hence the boundaries, which generally are narrow and well defined in oceanic crust, become broad and diffuse on the continents. The boundary between the Eurasian plate and the Indian plate, for example, is more than 2,000 kilometers wide in some of the places where the continental crust of one plate is in contact with the continental crust of the other.

Second, the continental crust is markedly nonuniform in its mechanical properties, consisting as it does of belts of older rock and other preexisting structures that can localize new episodes of deformation. A zone of deformation extending well into a continent may thus form structures so greatly influenced by the anisotropies in the crust that the structures are hard to relate to the tectonic activity at the border of the plate. In such a zone of deformation it is difficult to define a plate boundary: the entire zone functions as the boundary. Commonly the older rock assemblages and structures of the continents have been subjected over time to the activity of many different plate-boundary systems. Therefore the record in the rock is often fragmentary and difficult to interpret. Under these circumstances the study of modern plate-boundary systems can show how modern and ancient plate boundaries evolve. Some examples from the three types of plate boundaries can serve as an introduction to the more complex patterns that result from the superposition of several tectonic episodes.

The divergence of two plates along a divergent plate boundary that crosses the continental lithosphere begins as the crust and its underlying lithospheric mantle become stretched and attenuated. Crustal faults develop in long, narrow zones, and within these zones the faulted crustal rocks subside differentially, forming great tilted blocks. Since the upper part of the mantle participates in the stretching, material from lower in the mantle (the hotter, more ductile level called the asthenosphere) rises to take its place, increasing the heat flow through the lithosphere. The result is the partial melting of the mantle and a characteristic volcanism of basaltic rock that is often alkalic (that is, rich in sodium and potassium).

Sometimes the divergence ends after only a few tens of kilometers of stretching, so that the zones of attenuated continental crust remain rifted scars in the

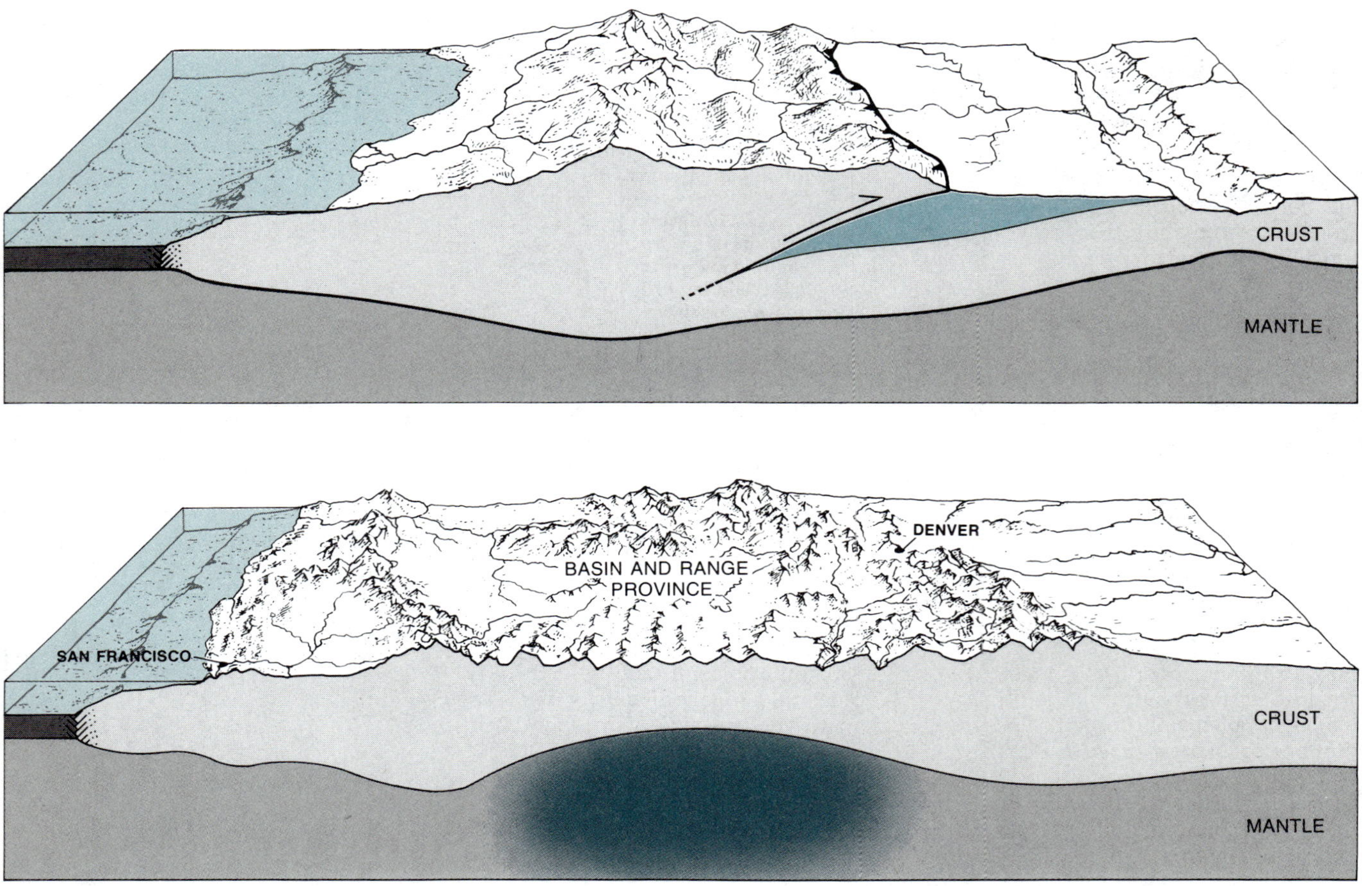

FLOTATION EQUILIBRIUM describes the relation of the continental crust to the underlying mantle. The crust, which is the lighter of the two, behaves as if it were floating; thus regions of great elevation, such as mountain chains, tend to be places where the crust is notably thick. Here two exceptions are shown. In the top drawing the crust near a mountain chain has flexed downward as if it were an elastic sheet supporting a weight. In the downward-flexed region, which has filled with sedimentary rock (*color*), the crust is thicker than one would judge from the elevation of the surface. In the bottom drawing, which shows the western U.S., a part of the mantle is hotter and so less dense than usual (*color*). Its buoyancy supports the crust, which is thinner than one would judge from the elevation of the surface.

continents. Some young, still active examples are the Rhine valley and associated rifts in central Europe, the East African rift valleys and the Rio Grande rift of the American Southwest. More ancient examples are the Oslo rift in southern Scandinavia, which is some 280 million years old, the Keweenawan rift of the central U.S., which is a billion years old, and the Athapuscow and Bathurst rifts in northwestern Canada, which are more than two billion years old.

In other instances the divergence continues. The attenuation of the crust thus becomes more extreme. It culminates in the separation of two bodies of continental crust and the formation of a new ocean basin, underlain by a widening span of oceanic crust. Each margin of the continental crust moves away from the region of divergence. The hot upper mantle moves with it. With time this hot upper mantle cools and contracts, causing the crust to subside. Meanwhile the faults that accompanied the stretching and attenuation of the crust become inactive. Sedimentary rocks begin to accumulate thick buildups above the thinned, subsiding crust. They also accumulate above the transition from continental crust to oceanic crust, forming the characteristic continental margin (called a passive continental margin) that flanks many ocean basins.

A profile through a rifted margin illustrates its evolutionary history. Thinned and faulted continental crust is overlain by a sequence of sedimentary and volcanic rocks deposited in faulted rifts during the initial phase of divergence. Those rocks are overlain in turn by a thick blanket of unfaulted sedimentary rock deposited during the latter, more gradual phase of subsidence. Well-studied examples of rifted margins are the Atlantic margins of the East Coast of the U.S. and the west coast of Africa. The margins generally have high temperature gradients during their early evolution, and so they are favorable sites for the maturation of organic matter into deposits of petroleum and natural gas. Occasionally the divergence of two continental bodies occurs near an older continental margin, and fragments of continent are rafted away to form small plateaus of continental crust partially or entirely submerged in the oceans and surrounded by oceanic crust. Examples include the Lord Howe Rise (with its highest part, New Zealand) in the southwestern Pacific and part of the Kerguelen and Mascarene plateaus in the Indian Ocean.

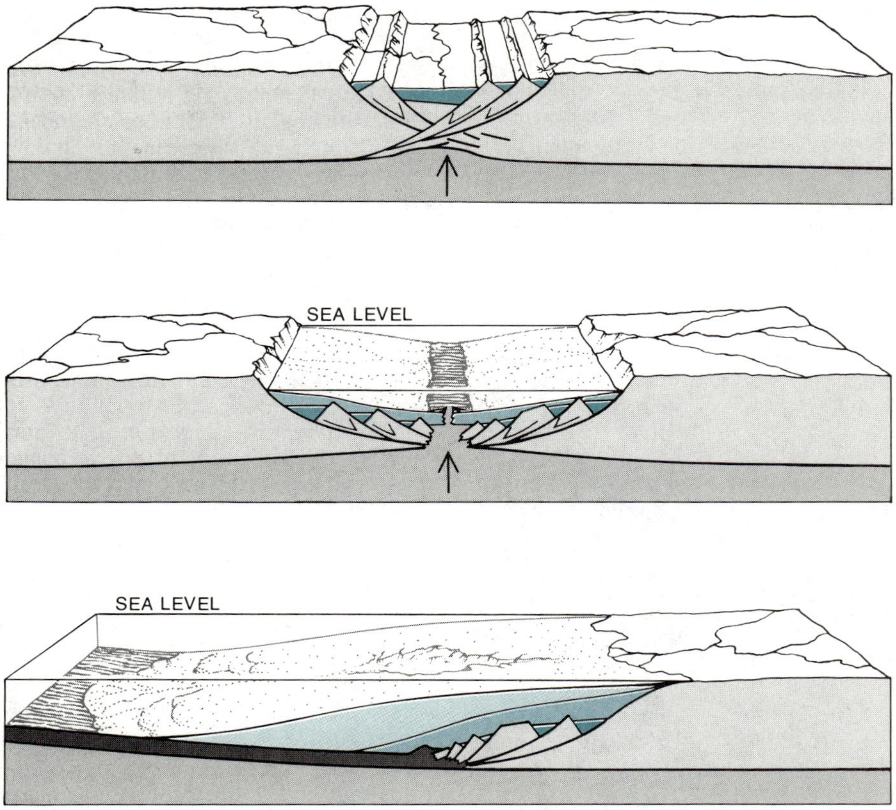

DIVERGENT PLATE BOUNDARY, where two plates move apart, causes extension and thinning of the continental crust. Initially (*top drawing*) the upper crust develops a series of brittle faults. Often the fault blocks rotate as they subside. The extension at deeper levels is less well understood. The subsidence (*middle drawing*) provides a site for the accumulation of nonmarine or shallow-marine sedimentary rocks (*medium colors*) and the rise of igneous rocks (*dark gray*). If the plates continue to diverge (*bottom drawing*), two continents result, and oceanic crust forms between them. The thinned margin of each continent subsides as it moves away from the zone of divergence; thus it is covered by unfaulted sedimentary rocks (*light color*).

At transform boundaries, where two plates slide horizontally past each other along vertical or nearly vertical faults, crust is neither created nor destroyed. The horizontal displacement along the length of the boundary can measure hundreds of kilometers, even thousands. When the boundary crosses continental crust, the displacement is commonly distributed across a zone of faults as much as several hundred kilometers wide. Preexisting belts are shifted laterally, and parts of them can be rotated, greatly disrupting their original continuity. Moreover, offsets or bends in the faults can give rise to local regions of divergence or convergence.

Here are two examples. The Alpine fault in New Zealand is part of a transform system along the boundary between the Pacific plate and the Indian plate. It passes through a fragment of continental crust that was rafted away from Australia about 100 million years ago. The horizontal displacement along the fault is now about 400 kilometers, but the movement of the plates is not limited to displacement. In addition rock assemblages and structures created by activity at older plate-boundary systems have been rotated and bent, recording a total of about 1,200 kilometers of differential motion. It can be shown that the motion was purely transform about 40 million years ago and later became oblique, with components of both transform and compression. The compression has thickened the crust and raised a chain of high mountains: the New Zealand Alps.

The Dead Sea fault zone in the Middle East is a transform system connecting a divergent plate boundary in the Red Sea to a convergent plate boundary in the Taurus Mountains of southern Turkey. In places the fault zone steps to the west, cutting across the direction of transform motion and thereby creating small regions where the transform motion causes the crust to stretch, become attenuated and subside. The Dead Sea, the Sea of Galilee and the Gulf of Aqaba are all instances of such "pull-apart basins" along the fault. North of the northernmost basin the fault zone bends and steps to the east, producing the opposite result: the compression and thickening of the crust, which has raised the Palmyran Folds. In this way some of the northward motion of the Arabian plate with respect to Europe has been absorbed by convergence and shortening within the continental crust.

It is convergent plate-boundary systems that generate most of the continental crust. Among the three types of boundaries they are the most complex type, and in addition they deform the continental crust across the widest region. In the most usual configuration of a convergent boundary system a plate of oceanic lithosphere is subducted under an overriding plate of either oceanic or continental lithosphere. At increasing distance from the zone of subduction the overriding plate commonly shows a sequence of geologic features: first an accretionary wedge of folded and faulted sedimentary rocks and fragments of oceanic crust scraped from the top of the downgoing plate; then a topographic maximum (an "outer-arc high") formed by the most elevated parts of the accretionary wedge; then a fore-arc basin, which accumulates sediment from the adjacent elevations, and finally a volcanic arc, the most characteristic feature, fueled by magma rising from the subducted plate and the mantle just above it. If the overriding plate is oceanic lithosphere, the geologic features form what is called an island arc. If it is continental lithosphere, they form a continental-margin volcanic arc.

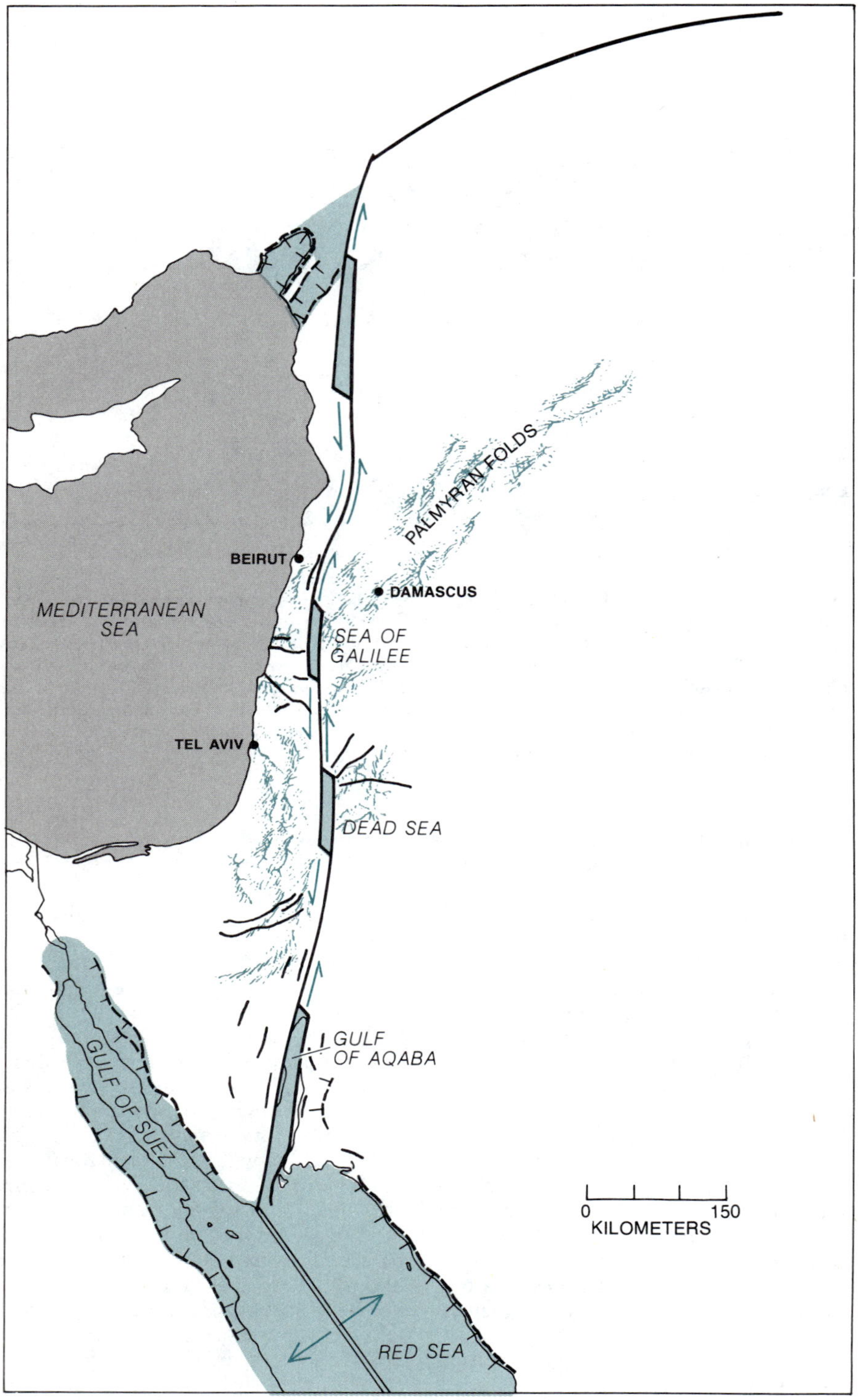

TRANSFORM PLATE BOUNDARY, where two plates slide past each other, is exemplified by the Dead Sea fault zone in the Middle East. The crust at the east of the fault is moving north with respect to the crust at the west, and the relative displacement, which amounts to about 105 kilometers in the southern part of the zone, has opened a number of gulfs and seas, of which the Dead Sea is one. In addition parts of crust have shortened to form the Palmyran Folds.

Behind the volcanic arc the overriding plate may be extended or compressed. Alternatively it may be relatively passive. If the plate is in extension, structures similar to the ones at divergent plate boundaries can form. A broad zone of stretching may thin the lithosphere and the crust, forming depressions such as the Aegean Sea of the eastern Mediterranean. If the extension of continental crust behind the arc proceeds until new oceanic crust is formed, a marginal sea will result. It will intervene between the greater part of the continental mass and the newly rifted fragment of continental crust, just as the marginal sea called the Sea of Japan intervenes between Asia and the islands of Japan. If the overriding plate is in compression, folds and faults arise in belts to accommodate the shortening and thickening of the crust behind the arc. In the Andes such features are found more than 800 kilometers behind the subduction zone.

At convergent zones new material from the mantle is added to the crust. In particular the subduction of oceanic lithosphere carries some ocean-floor sediments and the uppermost part of the oceanic crust downward into the mantle. The sediments and the crust contain water, and the water reduces the melting temperature of certain components of the subducted material. It also reduces the melting temperature of certain components of the mantle of the overlying plate. In short, the subduction of oceanic lithosphere causes partial melting at depth. The melted igneous material rises into the overlying rock. There it may cool and crystallize to form plutons: large subterranean igneous bodies. It may also reach the surface as lava or as explosive volcanic products such as pumice and ash. It is enriched in the elements common in continental crust; thus the partial melting advances the chemical differentiation of the outer part of the earth.

Geochemical studies of the igneous products show that they have had a complex, multistage history before coming to rest. In many instances the igneous rocks have been contaminated by contact with older crustal rock, so that not all their volume represents new material derived from the mantle. Indeed, some igneous rocks are derived entirely from the melting of continental crust; hence they add nothing to the volume of the crust. The question of how much of the material added to the continents through igneous intrusion is new and how much is recycled is still unresolved.

In any case the igneous intrusions increase the temperature within the lower crust, and the increase enhances the ability of the crustal rock to lose brittleness and deform in a ductile way. The structures in this ductile part of the crust may thus form large, complex folds. Moreover, the preexisting rocks may recrystallize into rocks with new mineral assemblages. The deformation and recrystallization may obscure and even obliterate the preexisting rock types and deformational patterns, so that it is difficult to elucidate the origin and evolution of the older rocks when erosion uncovers them at the surface of the earth. In general the regions of ductile deformation in the continental crust grade upward and laterally into regions of brittler deformation where the temperature remained lower.

Inevitably the convergence of plates leads to collisions between island arcs and continents. The arcs (along with oceanic islands and plateaus) are a transitional type of crust thicker and less dense than oceanic crust but not as thick and not as "light" as most continental crust. Nevertheless, the arcs, like continental crust, tend to resist subduction. Thus the elimination (by subduction) of the oceanic crust between an arc and a continent that are on converging plates leads to their becoming sutured together. The convergent motion during the collision may be perpendicular to the convergent boundary or it may be oblique to the boundary and have a component of transform motion. Where the motion is oblique the deformation within the collision system will have the characteristics of both convergent and transform boundaries.

The tectonic events at Papua New Guinea exemplify the geologic evolution caused by the collision of island arcs and continents. Here the convergence of the Australian plate and the Asian plate over the past 40 million years has driven one island arc or possibly two up over the edge of the Australian plate. The rock assemblages of the arc (or arcs) have been slivered and foreshortened, along with the assemblages of the ancient margin of Australia. Indeed, the northern edge of Australia has been shortened and thickened as far as 300 kilometers from the site of the collision. The disruption of the arc and of the continent has therefore been severe, but not so severe that the original relations among the rock assemblages cannot be deciphered.

Another type of geologic evolution is found where two continents collide along a convergent plate boundary. Such collisions are occurring today along the Alpine-Himalayan chain, where the Indian plate, the Arabian plate and the African plate are each colliding with the Eurasian plate. In the eastern Mediterranean the collision zone is more than 500 kilometers wide. There the geologic evidence suggests that several small fragments of continental lithosphere were swept together between the converging plates. The "buoyant" fragments stayed at the surface while oceanic tracts were subducted. The continued convergence in the region has deformed both the fragments and the margins of the plates, so that the collision system now extends across a broad zone.

One of the characteristics of collision systems, particularly those between continents, is subhorizontal decoupling, a process in which crustal sheets 10 to 20 kilometers thick slide over one another for tens or hundreds of kilometers. Such displacements stack and thicken different parts of the crust into a series of irregularly deformed and folded sheets, so that the rock assemblages and structures at depth in the crust cannot be pre-

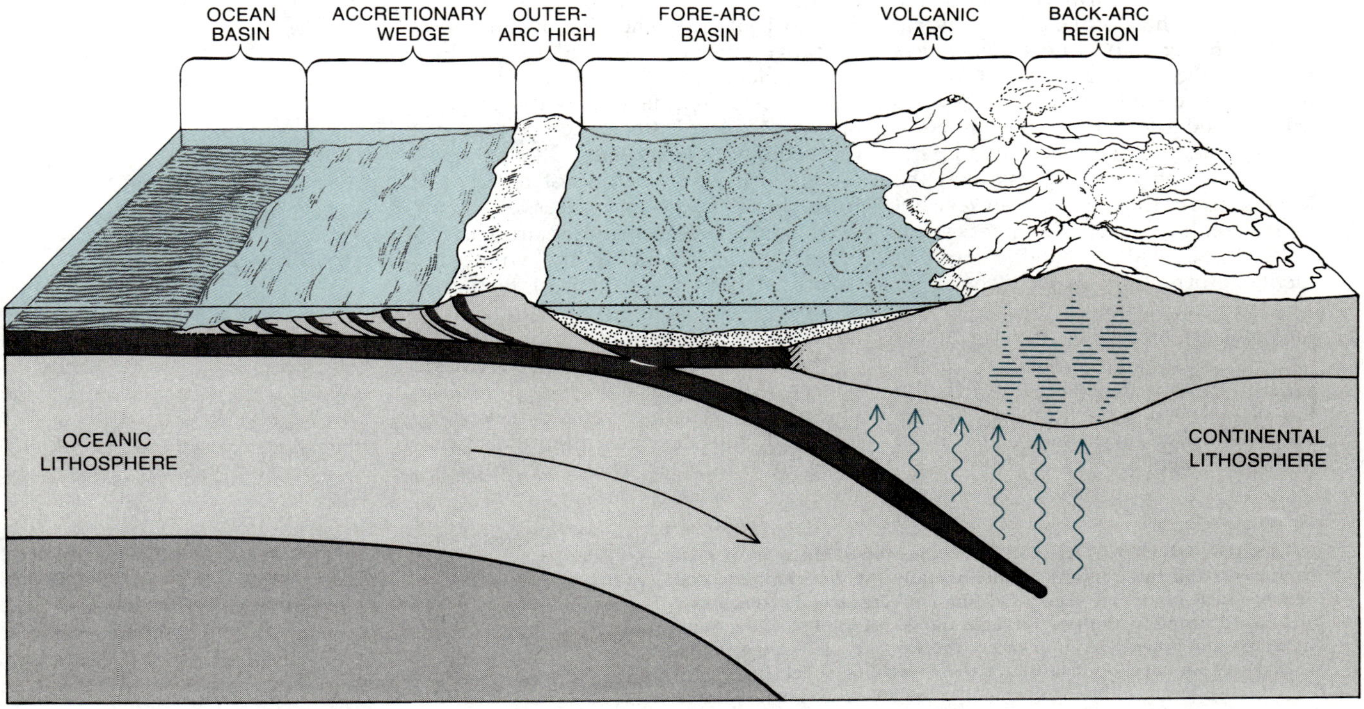

CONVERGENT PLATE BOUNDARY, where two plates collide, is marked by a characteristic sequence of geologic features in the overriding plate. In the most usual configuration oceanic lithosphere (crust and upper mantle) is subducted under continental lithosphere. Fragments of oceanic crust and sedimentary rock scraped from the subducted plate form an accretionary wedge and an outer-arc high. Next comes a fore-arc basin, which accumulates sediments from the adjacent elevations, and then a volcanic arc, the most characteristic feature resulting from subduction. Some of the magma rising from the subduction zone solidifies in the crust. The back-arc region behind the volcanoes may show convergence (such as crustal faulting and folding) or divergence (such as crustal thinning and subsidence).

dicted from the assemblages and structures exposed at the surface. Another characteristic is that the convergence, which typically occurs along irregular boundaries in crust that is very anisotropic, causes complex motions of small crustal fragments within the convergent system. These local motions may be divergent, transform or convergent.

The most spectacular example of a convergent plate-boundary system active today is in Asia, where Peter Molnar of the Massachusetts Institute of Technology and Paul Tapponnier of the French National Scientific Research Council (C.N.R.S.) were the first to recognize that deformation extends across a region 3,000 kilometers wide. The collision some 50 million years ago between the Indian plate and the Asian plate represented a collision between the continental lithosphere of India and that of Asia. Since then a continued convergence that may total more than 2,000 kilometers has been absorbed principally by strain in the Asian plate. Broadly speaking, the Asian plate has absorbed the massive intracontinental deformation by compressional, transform and extensional faulting along young belts of deformation that generally follow older belts resulting from the activity of more ancient plate-boundary systems. To put it more simply, Asia has shortened longitudinally and extended latitudinally to accommodate the northward movement of India.

Meanwhile the northern edge of the Indian plate has broken into several gently dipping slabs whose pileup has thickened the crust and formed the Himalayas. The faulting in Asia extends nearly 3,000 kilometers from the collision boundary. Igneous activity in parts of the collision zone suggests that the deeper parts of the crust remain very hot today, creating an environment where rocks are recrystallizing and undergoing ductile deformation. Hence the rock assemblages and structures formed at earlier times are now being "overprinted."

From studies of the youngest deformed belts, such as the Alpine-Himalayan belt, it is plain that plate boundaries evolve rapidly. Island arcs can be created, travel thousands of kilometers and collide with continents in only a few tens of millions of years. Small continental fragments can be rifted and collide with continents over similarly short spans of time. Thus the deformed belts that make up the underpinning of the continents represent a long, complex history of superposed plate-boundary systems.

The end of such a history probably comes when a region gets to be so remote from plate-boundary activity that it is no longer under its influence. In many instances this happens when an ocean basin is finally closed by a continent-continent collision. Parts of the Ural Mountains of the central U.S.S.R. are an example. First, about a billion years ago, two continental masses were rifted apart and a large ocean basin was created between them. The subsequent closing of that ocean generated island arcs, which collided with the continents at various times until the sequence was ended by the collision of two continents 250 million years ago. Much of the deformed belt resulting from that collision (namely the chain of the Urals) lies far enough from any younger plate-boundary activity for it to have suffered no further deformation.

This is not to say the entire belt is immune. Rifting in the south has created oceans younger than 250 million years, some of which have closed and created new mountain belts; hence the southern extension of the Urals has been overprinted by the Alpine-Himalayan belt. Moreover, the northern extension of the Urals reaches the Arctic Ocean, where future plate-boundary activity may be in store. Similar histories can be read in the geology of essentially all the continents in the crosscutting of older deformed belts by younger ones.

Of course, some histories are easier to read than others. In the youngest deformed belts the timing of tectonic events can be distinguished with an accuracy of better than one million years. In the older belts the timing becomes poorer with increasing age. The relative timing of many events can be determined, but the contemporaneity of events over large areas is difficult to es-

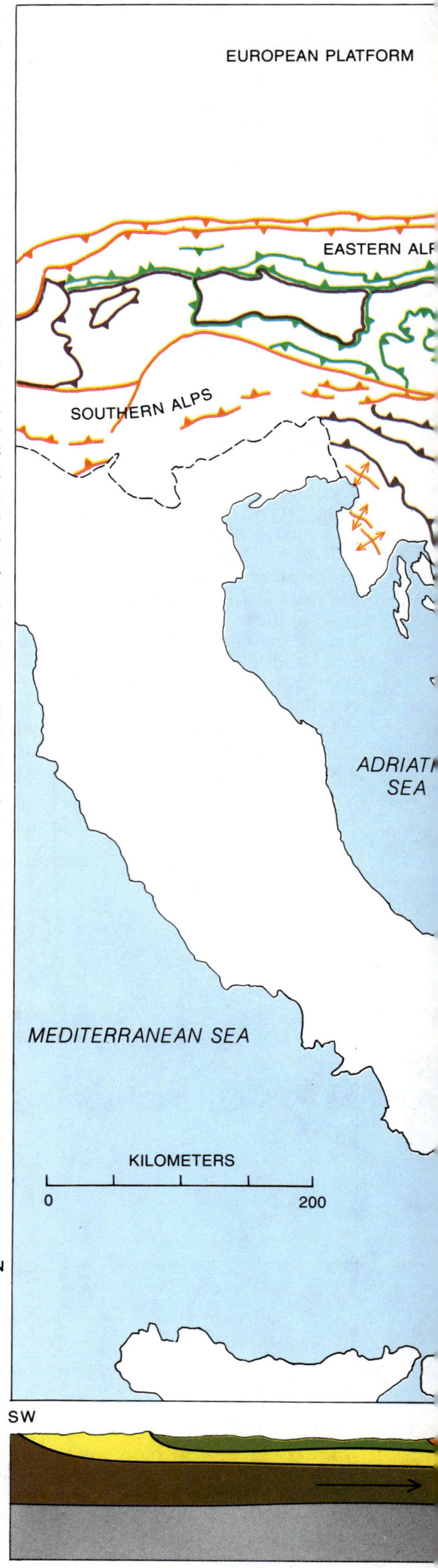

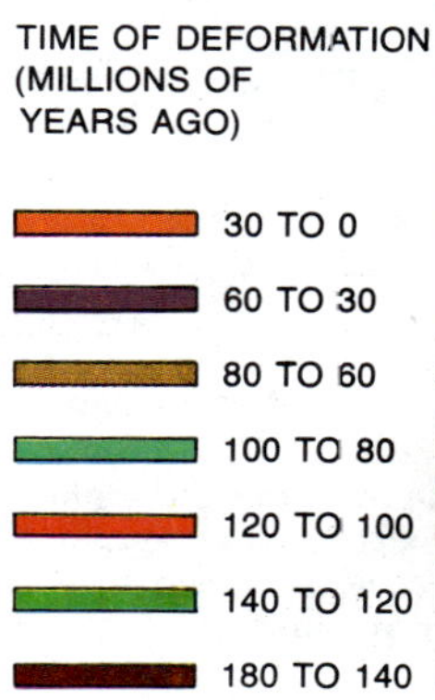

COLLISION OF CONTINENTS raised the mountain chains of eastern Europe and the Balkans. Fundamentally the African plate and the European plate converged, and the convergence shortened and thickened Europe by faulting its rocks into sheets and stacking them one above the other. The map shows the varying ages of the faults. Small barbs on each fault line face the overriding sheet. In general the stacking has a bilateral symmetry: the overriding sheets were driven northward in the northeastern half of the map and southward in the southwestern half. In the Pannonian Basin the crust has extended, thinned and foundered, so that the stacking is covered by sedimentary rock (*gray*). Broken lines indicate outcrops of deeper rock. The cross section suggests the complexity of the stacking; the colors employed in the section simply differentiate sheets of rock. The sheets are highly deformed; thus no individual sheet can be followed the width of the deformed belt resulting from the convergence.

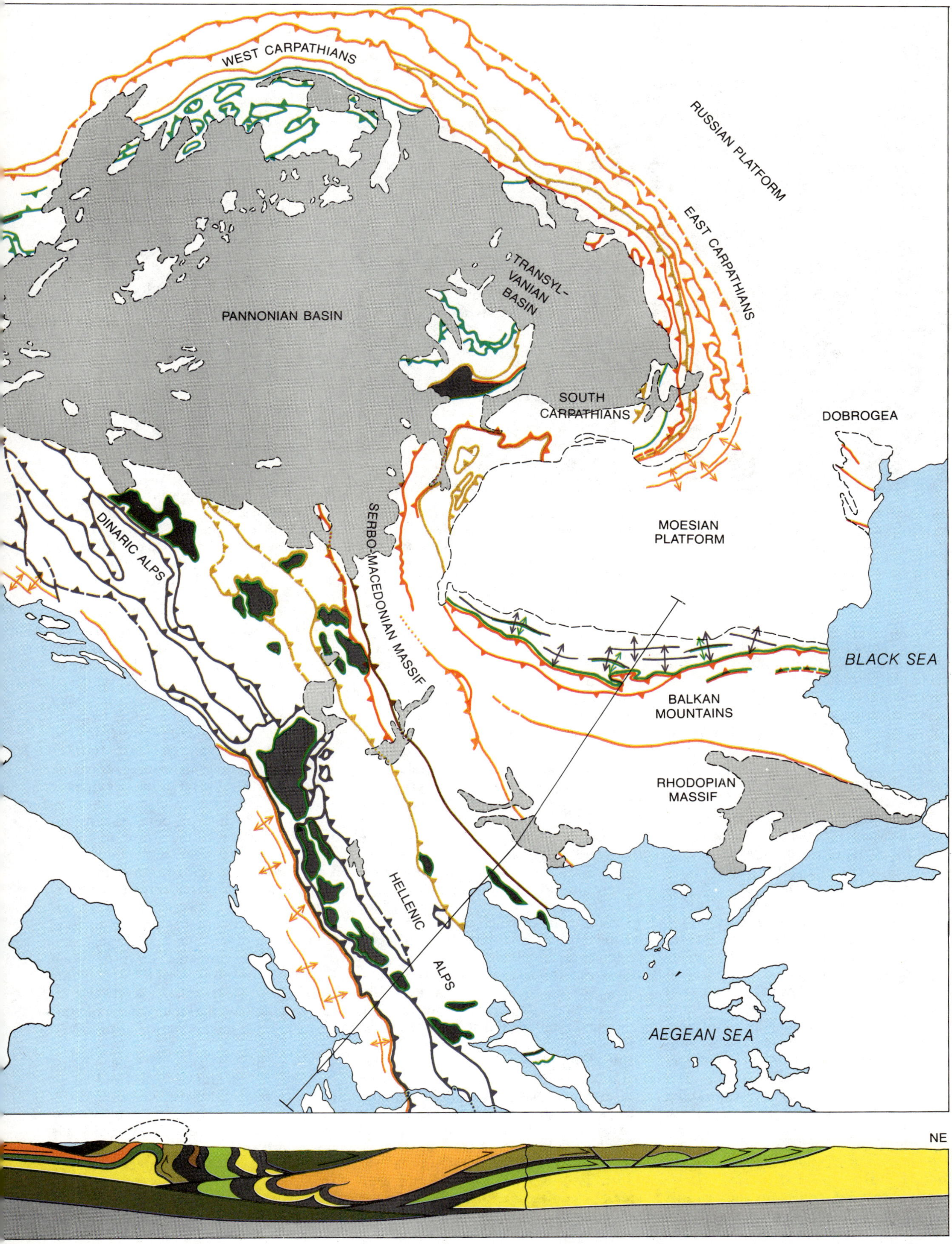
WEST CARPATHIANS
RUSSIAN PLATFORM
EAST CARPATHIANS
TRANSYL-
VANIAN
BASIN
PANNONIAN BASIN
SOUTH
CARPATHIANS
DOBROGEA
DINARIC ALPS
SERBO-MACEDONIAN MASSIF
MOESIAN
PLATFORM
BLACK SEA
BALKAN
MOUNTAINS
RHODOPIAN
MASSIF
HELLENIC
ALPS
AEGEAN SEA
NE

BRITTLE DEFORMATION is characteristic of the rocks at shallow levels in a belt deformed by plate convergence. Hence sheets of rock tend to be thrust one over the other for tens or hundreds of kilometers. This photograph was made in the Spring Mountains of southern Nevada. The dark rocks are Cambrian limestones that are 550 to 500 million years old. They have been thrust from right to left over Jurassic sandstones lighter in color, which are 200 to 175 million years old. A well-defined thrust fault some 30 kilometers long marks their interface.

tablish. Thus it becomes a challenge to make accurate reconstructions of ancient plate-boundary systems. In addition the older belts are likelier to be ones where the once continuous locus of deformation has been disrupted. For example, the continuations of the ancient deformed belts in Australia are now found in India, Africa, Antarctica and South America.

The erosion of old deformed belts in regions where convergence has thickened the crust offers a valuable opportunity to study rocks that were once at deep crustal levels. It is remarkable just how deep erosion can reach. Erosion cuts into the terrain (a process taking tens of millions of years, even several tens of millions), progressively uncovering deeper levels, and since the crust is "buoyant," the removal of material from the top of it causes the remaining crust to rise. In effect it is timber pushed toward a saw. In this way rocks from depths as great as 30 or 40 kilometers come to be exposed at the surface.

By the time the rocks are exposed the plate-boundary forces that shaped them have been long inactive. Still, the examination of the rocks enables geologists to infer the processes, the temperatures and the pressures that existed as the rocks deformed, and from such results a picture of the three-dimensional response of the continental crust to plate-boundary activity can be constructed. Only the deepest crustal levels escape exposure by erosion. Studies of inclusions in igneous rocks, chemical studies of igneous rocks and geophysical studies suggest that much of the deepest crust has a composition not unlike the composition of the shallower crust, except that the deep rocks are recrystallized so that their mineral assemblages are ones that are stable at high temperatures and pressures.

Some of the older deformed belts in the continents, including most of the Archean belts (the belts from 2.5 to 3.8 billion years old), have been difficult to interpret as plate-boundary systems analogous to modern ones. To be sure, the types of rocks in the older belts are similar to the types found in modern convergent systems. Their arrangements and structures, however, are somewhat different. Typically the older belts consist of bodies of volcanic and sedimentary rock, irregular or elongated in shape, along with large expanses of intrusive granitic and deformed metamorphic rocks that include mineral assemblages formed at high temperatures and pressures. The volcanic and sedimentary bodies resemble those found in modern island-arc settings. The only notable differences are that basaltic rocks are more abundant in the older belts and that some of the basalts are richer in magnesium and poorer in silicon than most modern basalts. On the other hand, many features of the younger belts appear to be missing in older ones. Thick, widespread sequences of shallow-water sedimentary rocks of the type that develop on modern passive continental margins are one example. Widespread vertical stackings of sheets of crustal rock that have been thrust one over the other are a second example.

Investigators of the Archean belts have taken two points of view: that the Archean belts are the result of plate motions whose geometry and intensity differed from those of modern plates, or that plate tectonics did not operate in the Archean, so that a mechanism of geologic evolution not observed today must be invoked instead. Although each view has its adherents, a modification of plate-tectonic theory can also be proposed. Perhaps the volcanic and sedimentary bodies in the Archean belts represent island arcs and their associated fore-arc basins and marginal seas, all of which were swept against small colliding continental nuclei. Larger, continental masses developed progressively, so that by about 2.5 billion years ago some orogenic belts began to take on a more modern appearance.

The rate at which continental crust has been formed remains a major question. On the one hand, the examination of orogenic belts indicates that many of them include a large amount of material derived from older belts by reworking or remelting. In addition many belts include rocks derived directly from the mantle by volcanic activity. Further still, many belts incorporate oceanic lithosphere. On the other hand, evidence suggests that continental material may sometimes be lost by being taken into the mantle. The studies show that orogenic belts vary widely in the balance of these processes. For example, the orogenic belts that were formed in north-central Canada between 2.5 and 1.8 billion years ago are proving to include much crust older than 2.5 billion years. In contrast, an orogenic belt of the same age in the southwestern U.S. contains little if any older crust. The studies do seem to indicate that the volume of continental crust has increased with time.

One of the very oldest Archean belts, about 3.8 billion years old, is in the continental crust of Greenland. Its rocks are sedimentary and igneous, and material in the sedimentary rocks has been derived in part from some older continental rocks. No direct evidence from that earlier era has been discovered, however, and so there remains a gap of 800 million years beginning 4.6

billion years ago, when the solar system is thought to have formed. In the later part of this gap the moon was intensely cratered; surely the earth was subjected to a similar bombardment. No signs of it remain. They were probably erased by the dynamic processes that have continuously created and reworked the continental crust.

Clearly the earth is an evolving body whose distribution of heat controls the motions, thickness and ductility of the lithosphere and the generation of igneous and metamorphic rocks. The generation of heat by radioactivity in the earth was probably about three times greater in the Archean than it is today. Temperature gradients in the earth were probably greater too, and that may help to explain at least some of the differences between Archean belts and younger ones. Until the variables that affect the formation of orogenic belts are better understood, the Archean belts will remain a major challenge to the understanding of how the continents came into existence.

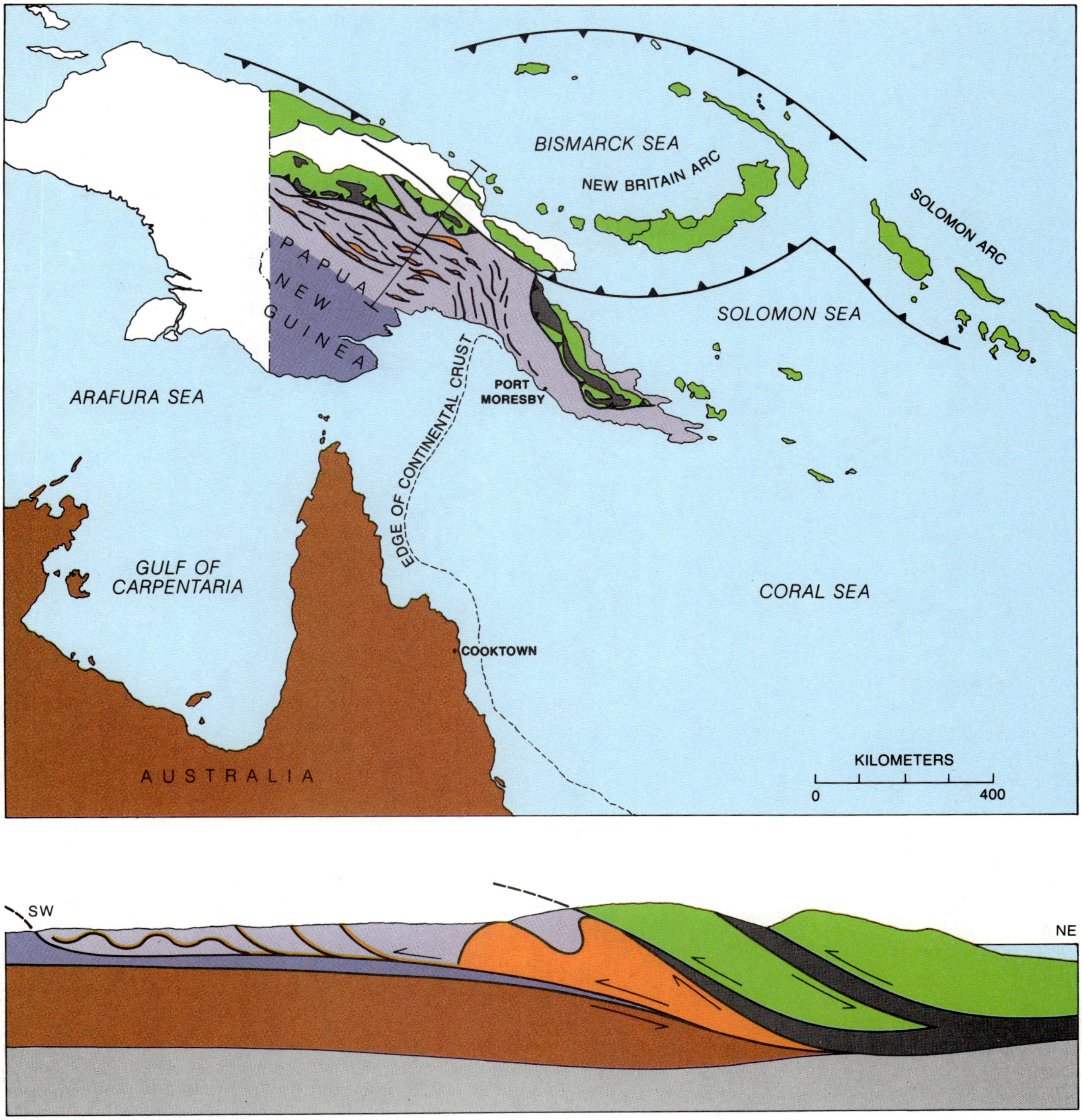

COLLISION OF ISLAND ARCS with the northern part of the Australian continental crust over the past 40 million years has sutured some arcs (*green*) and also some remnants of oceanic crust (*gray*) onto the northern part of Papua New Guinea. The arcs were probably formed by the subduction of the Australian plate under the Pacific plate; then the convergence of the plates carried Australia into the zone of the subduction. The continued convergence has now folded and faulted sedimentary rocks along the margin of Australia (*purple*) and also the ancient crust of Australia itself (*orange*). Darker shades signify rock that escaped deformation. The cross section is schematic.

6

THE OCEAN

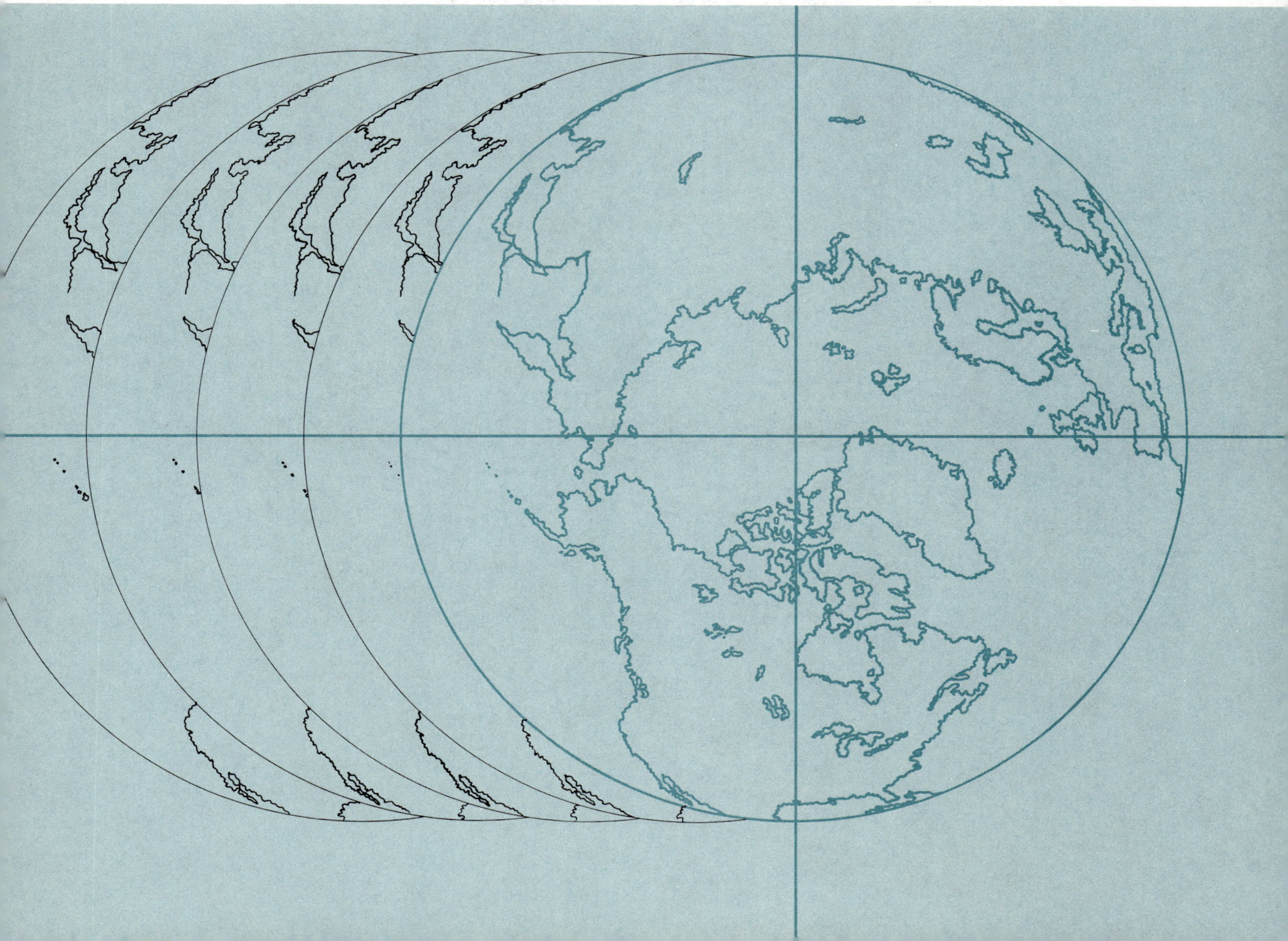

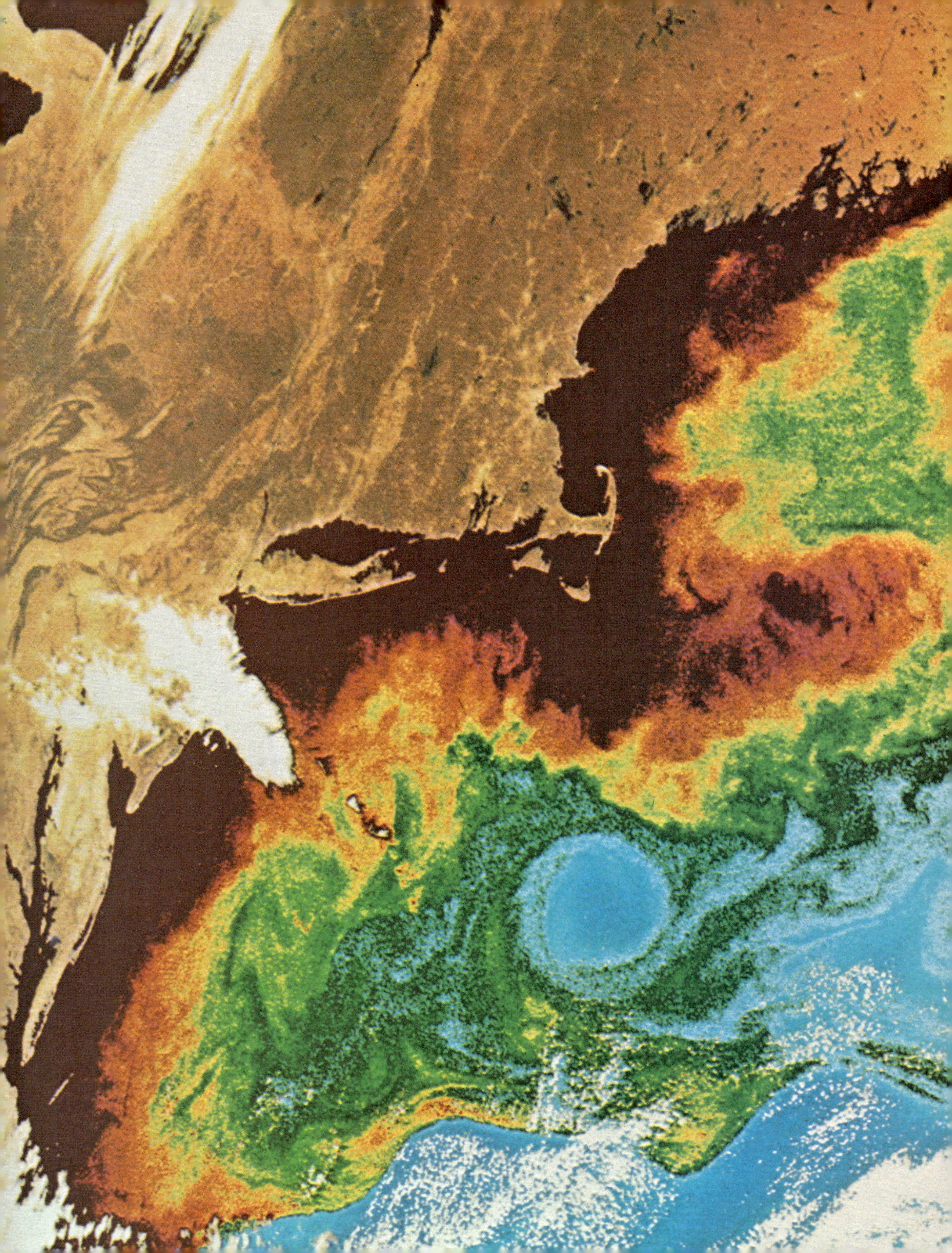

by WALLACE S. BROECKER

The chemistry of the ocean, whose constituents interact with those of the air and the land to support life and to influence the climate, is now known to have undergone changes since the last glacial epoch

Water and air interact with the lithosphere to provide chemical support for the community of living things. The ocean and the atmosphere constitute a single system that functions as an integrated chemical plant. In attempting to reconstruct the history of the earth, geologists have tended to think of the chemical composition of the ocean and the atmosphere as being essentially static through time. The reason is that geochemists have been slow to come up with reliable paleochemical indicators.

In the past 10 years there have been important breakthroughs in the study of ocean chemistry. There is now reasonably firm evidence, at least in the case of those constituents of the seas' salts that are heavily incorporated into organisms and processed by them, for significant changes in concentration over time. Some of these fluctuations (the best-documented ones) seem to have been in tune with major cycles of glaciation that have dominated the earth's environment for the past million years or so. Others were surely related to changes in the architecture of continents and ocean basins caused by movement of the great lithospheric plates. Still others may be effects of such geologic catastrophes as the great asteroid or comet impact that, it now appears, marked the end of Cretaceous time.

Fluctuations in the ocean's chemistry are worth studying for what they tell about variations in the planet's environment in the past. Ocean chemistry has affected the environment primarily by driving changes in the atmospheric content of one constituent in particular: carbon dioxide (CO_2). Much of the solar energy impinging on the earth is reradiated from the surface in the form of infrared radiation, which is strongly absorbed by carbon dioxide and so is trapped in the lower atmosphere as heat. The partial pressure of carbon dioxide in the atmosphere therefore influences temperature and hence also rainfall, winds and the extent of ice cover. It is well known that man's activities, notably the clearing of forests and the burning of fossil fuels, are increasing the atmosphere's carbon dioxide content, but only recently has it become clear that there were carbon dioxide changes in the past as well.

The major immediate source for the carbon dioxide in the air is the ocean, which at present holds, dissolved in one form or another, 60 times as much carbon as there is in the atmosphere's carbon dioxide. The transactions controlling the partitioning of carbon between the sea and the air depend on the detailed chemistry of those constituents of sea salt that are heavily utilized by marine organisms. To know the ocean's role in the carbon cycle it is necessary to study mechanisms controlling the concentrations of such constituents in the ocean: the interconnected feedback loops whereby those concentrations are driven toward dynamic equilibrium. Somehow the ocean must cope with the heavy demand of marine organisms for these critical ingredients, maintaining a balance between input from either erosion or volcanic activity and ultimate loss by burial in sediments.

The control mechanisms can best be understood if one thinks of the ocean as a giant two-reservoir aquarium. The upper reservoir interacts with the air and the land. Water evaporates from the sea surface, condenses and falls onto the land as rain. The percolation of rainwater through the land dissolves and carries into the ocean some of the seawater constituents I shall be discussing. Others are supplied by volcanic activity, both on the land and on the sea floor. The upper reservoir is lighted by the sun, whose energy is converted by plants (mainly microscopic phytoplankton) into organic matter; the plant matter supports a food chain of animals and bacteria. Debris from these organisms falls from the upper reservoir into the dark lower reservoir, where it nourishes other animals and bacteria. Some residues fall to the ocean floor, where they are decomposed by worms and bacteria. A small fraction of the residues remain unconsumed and are buried, and so the material dissolved from the land or injected into the atmosphere or the ocean as volcanic gas is ultimately deposited as sediment.

To keep things simple I shall deal here only with dissolved oxygen, calcium ion, phosphate and three carbon compounds: carbon dioxide, carbonate ion and bicarbonate ion. The choice of these constituents reflects the chemical requirements of marine organisms. Marine animals need dissolved oxygen for respiration. For their biochemical functions all organisms need carbon and phosphorus (along with hydrogen and oxygen, which are of course superabundant in water). Some organisms house themselves in chambers made of calcite (calcium carbonate).

Calcium enters the aquarium in the form of the doubly positively charged calcium ion (Ca^{++}) dissolved in river water and is lost to sediments as calcium carbonate ($CaCO_3$). Carbon enters through the erosion of carbonate miner-

CHLOROPHYLL CONCENTRATION in waters off the U.S. and Canadian coast is mapped in the false-color image on the opposite page. The nearly lifeless warm water of the Gulf Stream (and of a circular eddy pinched off from it) is bright blue, and successively increasing chlorophyll concentrations are represented in light blue, green, yellow, reddish brown and dark brown. Chlorophyll is manufactured by phytoplankton, the microscopic marine plants at the base of the oceanic food chain. As is described in this article, these microorganisms have a major role in the control mechanisms that regulate the chemistry of the ocean-atmosphere system. Data for the image were collected by the coastal-zone color scanner carried aboard the satellite *Nimbus 7*. It records energy radiated from surface waters in spectral bands that are more or less strongly absorbed by chlorophyll. The experimental team for the color scanner is headed by Warren A. Hovis, Jr., of the National Oceanic and Atmospheric Administration.

als (as the doubly negatively charged carbonate ion, CO_3^{--}) and from volcanic sources (as carbon dioxide gas); it is lost both as calcium carbonate and as organic matter in soft-tissue residues. Phosphorus is supplied as phosphate (PO_4) by the erosion of phosphate minerals and is lost to sediment in soft-tissue residues. Dissolved oxygen (O_2) is generated by plants and consumed by animals. In this simplified aquarium world four control loops are needed to keep the loss of ingredients in balance with their supply. A fifth control loop is needed to maintain electrical neutrality by balancing positive and negative ions.

I shall not deal with nitrogen, even though it is a key limiting ingredient for organisms. Two justifications for this simplifying omission are the superabundance of dissolved nitrogen gas and the observation that the ratio of nitrate (NO_3) to phosphate in seawater is everywhere about the same. Apparently organisms that are able to convert nitrogen gas into nitrate (the form most readily utilized by most plants) manage to maintain in the sea a nitrate-to-phosphate ratio that meets the biochemical demand. Since there is no such means by which living things can enhance their supply of phosphorus, it is phosphorus that becomes the ultimate limiting ingredient.

One must also consider the "residence time" of each of the ingredients in the ocean, which is to say the ratio of its total inventory in the ocean to its rate of supply (or of loss). In today's ocean the residence time, thus calculated, for phosphorus atoms is about 100,000 years, for carbon atoms about 165,000 years and for calcium atoms about a million years. The sizable difference between the residence time of phosphorus and carbon on the one hand and of calcium on the other allows for further simplification when one considers chemical changes associated with glaciation. Individual glacial episodes average some 50,000 years in duration. On such a short time scale the calcium content of sea salt cannot change significantly. Oxygen does not lend itself readily to a residence-time calculation, but there is so much of it in the atmosphere that its abundance is not likely to change significantly even in a million years. In considering chemical changes since the last glaciation, therefore, one needs to deal only with carbon and phosphorus (and charge balance).

Finally, the mechanisms of water transport between the two reservoirs of the ocean must be taken into account. The two reservoirs are separated by the main thermocline: the boundary region separating the cold waters (maintained by the sinking of dense water from the surface in polar regions) from the warm surface waters (maintained by sunshine). The strong difference in density between the two realms limits the rate at which they mix, so that the mean time spent by a water molecule in the cold-water compartment between trips to the warm surface ocean is about 1,000 years. A typical phosphorus atom therefore makes about 100 trips back and forth across the thermocline in the course of its 100,000 years in the ocean. A carbon atom makes 165 such trips in the course of its residence time, a calcium atom 1,000.

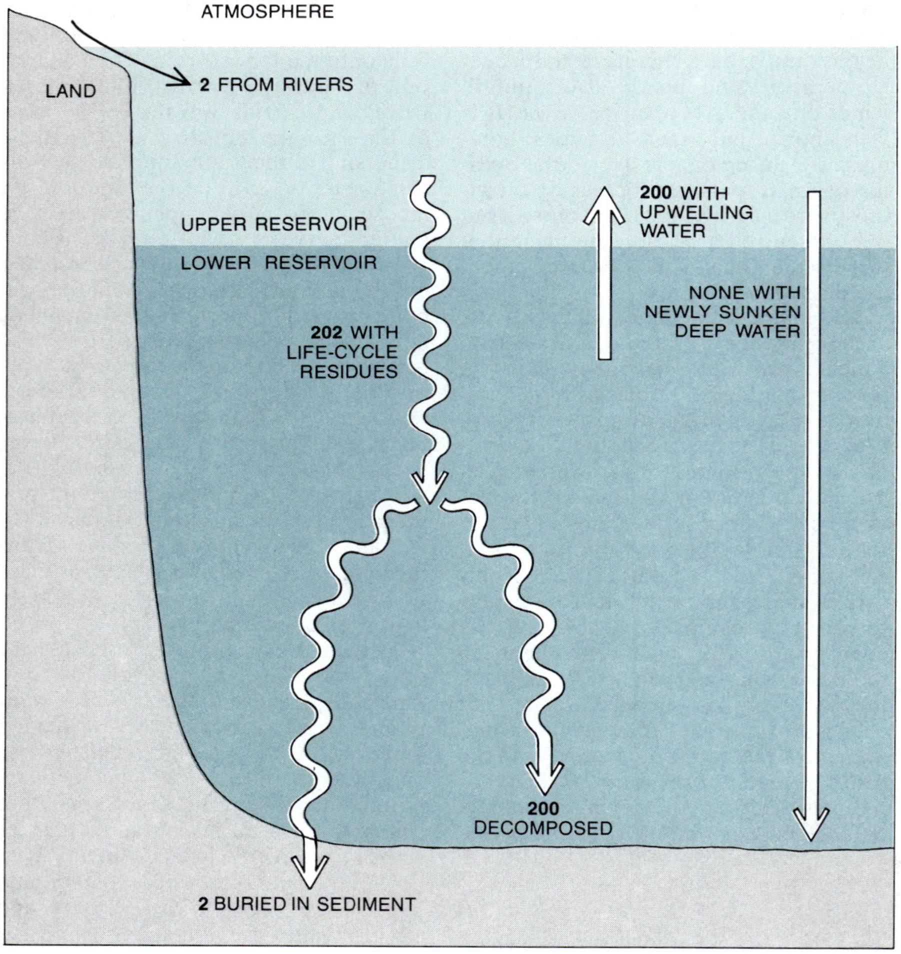

TWO-RESERVOIR OCEAN consists of a warm surface compartment and a much larger body of denser, colder water. The movement through the ocean of chemicals that are critical to marine life is represented here by an idealized phosphorus cycle. Phosphorus eroded from the land is delivered to the sea by rivers; eventually it is removed by being buried in sediment as part of undecomposed organic residues. Loss matches input (shown here in tens of billions of moles per year). Almost all the phosphorus reaching the warm surface ocean is taken up by marine plants and carried to the deep sea with falling residues, most of which decompose, releasing phosphorus into solution; the regenerated phosphorus returns to surface with upwelling water.

Although carbon and phosphorus move with upwelling water from the deep ocean to the surface layers, in their return trip to the deep sea these elements have a different option: they can fall, chemically bound within a fragment of organic matter. Almost all the phosphorus that moves up across the thermocline in upwelling water falls back to the cold-water compartment in organic particles, whereas only about an eighth of the upwelled carbon takes the particulate short cut; the remainder returns dissolved in sinking water. The reason for the discrepancy is that the ratio of carbon to phosphorus in sea salt is some eight times higher than the ratio in organic debris, so that in exhausting the available phosphorus the community of living organisms in the upper ocean uses up only 12.5 percent of the dissolved carbon. In removing carbon to form their tissues marine plants reduce the partial pressure of carbon dioxide in the surrounding water and thus also in the atmosphere. The effect is so large that if life in the ocean were suddenly suspended, the carbon dioxide pressure would rise significantly in surface water, and it would about triple in the atmosphere.

The history in the ocean of an average phosphorus atom can be traced from when it issues from the river mouth and enters the surface sea. Soon it is taken up by a plant. After being traded several times among the plant, animal and bacterial populations of the surface water, it is trapped in a residue (such as a fecal pellet) that falls into the deep sea. Of 100 phosphorus atoms bound in falling residues 99 are released back into solution; one is removed from the sea to permanent storage in the sediment. The redissolved atoms eventually find their way back to the surface and repeat the cycle. The average phosphorus atom is therefore dissolved in the deep sea in 100 separate 1,000-year terms, which are separated by brief interludes of only a few months during which it serves the biochemical needs of organisms in the warm surface ocean.

A match between the removal of phosphorus (with undestroyed residues) and phosphorus input to the sea (in river water) is maintained by adjustments in the ocean's phosphorus content. The higher the phosphorus content is, the more atoms reach the surface with upwelling water, the more residues fall from surface water and the more residues are sequestered in accumulating sediment. If some perturbation were to cause the loss of phosphorus to exceed input, the phosphorus inventory of the ocean would begin to decrease; there would be corresponding decreases in the rates of production and burial of residues. The mismatch between output and input would gradually be reduced until it eventually disappeared. The time scale for the adjustment would be of the same magnitude as the oceanic residence time of phosphorus atoms, or about 100,000 years.

How might the system be perturbed? Suppose a climatic change were to double the rate of exchange of water between the surface and the deep ocean. The delivery rate of phosphorus to the surface would double, as would the production and burial rates of residues. If the situation persisted, the ocean's phosphate content would decrease until it reached half of the initial value and the balance between input and output would have been restored. If, on the other hand, the river-input value were halved, the ocean's phosphorus inventory would eventually be halved. And if there were a 1 percent reduction (from 99 percent to 98) in the efficiency with which residues falling into the deep sea are destroyed, releasing phosphorus to dissolution, the phosphorus inventory would also be halved.

In my idealized ocean phosphorus is the limiting ingredient for life, and once its concentration is fixed so too is the rate at which residues are generated to rain down from the surface sea. This must be roughly correct, but ecological factors could alter the ratio of carbon to phosphate in soft-tissue residues or the ratio of falling calcite to falling soft-tissue residues. Putting such factors aside, one can assume that once the rate at which phosphorus rains down from the surface is set, so too is the total rate at which carbon rains down (in both soft tissues and calcium carbonate). At present, however, only about a sixth of the carbon delivered to the ocean by erosion and volcanism is lost with soft-tissue residues. The remaining five-sixths must leave the sea as calcium carbonate. How is the flow of this carbon regulated?

Whereas for phosphorus a balance between loss and gain is achieved by regulation of the rate of production of phosphorus-bearing residues, the balance for carbon is controlled by the dissolution of calcium carbonate residues. To understand this scheme of control by dissolution it is necessary to understand how the distribution of carbonate ion and its saturation level vary with depth. The carbonate content decreases with depth because carbon dioxide is removed from surface water as a result of photosynthesis (increasing the carbonate-ion concentration) and is returned to solution in deep water (decreasing the carbonate concentration). The saturation level for carbonate, on the other hand, increases with depth because calcite's solubility increases with pressure.

	INPUT TO OCEAN		SEA SALT	MARINE-ORGANISM DEBRIS	LOSS TO SEDIMENT	RESIDENCE TIME (THOUSANDS OF YEARS)
PHOSPHORUS	(EROSION)	1	1	1	(ORGANIC MATTER) 1	100
CARBON	(CARBON DIOXIDE, VOLCANISM)	100	1,000	125	(ORGANIC MATTER) 100	165
	(CARBONATE ION, EROSION)	500			(CALCIUM CARBONATE) 500	
CALCIUM	(EROSION)	500	5,000	25	(CALCIUM CARBONATE) 500	1,000

RATIOS of carbon and calcium atoms to phosphorus atoms are given for inputs to the sea, average sea salt, marine organisms and marine sediments, along with the average length of time atoms of each element reside in the ocean before being buried in sediment. A distinction is made between carbon added by volcanism and added by erosion of calcium carbonate ($CaCO_3$) and between carbon buried in sediments as organic matter and buried as calcium carbonate.

	WARM SURFACE WATER	AVERAGE DEEP WATER
PHOSPHATE	0	2.2
TOTAL DISSOLVED CARBON	2,000	2,275
BICARBONATE ION	1,764	2,140
CARBONATE ION	223	90
CARBON DIOXIDE GAS	13	45
CALCIUM ION	10,000	10,055
CADMIUM	0	.7
CARBON 13 / CARBON 12 (COMPARED TO STANDARD RATIO)	1.0024	1.0002

COMPOSITION of warm surface water is compared with that of average deep water. The concentration of the first six constituents is given in millionths of a mole per liter and that of cadmium in billionths. The ratios of carbon 13 to carbon 12 are compared with the standard ratio, in a National Bureau of Standards reference calcium carbonate. The differences between surface- and deep-water values observed for both chemical and carbon-isotope composition arise because organisms take up constituents in surface water; their soft-tissue residues and calcite ($CaCO_3$) shells fall into the deep sea, where decomposition and dissolution return the constituents to solution. A balance of electric charge is maintained: the sum of the charges contributed by positive calcium ions and by negative bicarbonate and carbonate ions remains constant.

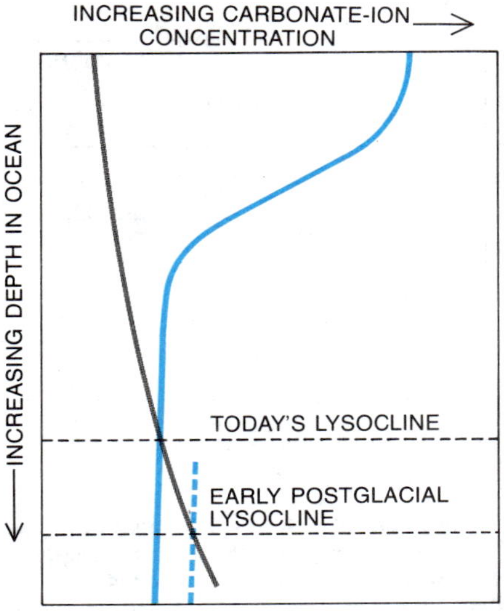

CARBONATE CONTENT of ocean water decreases with depth (*colored curves*), whereas the saturation content increases with depth (*gray curve*). The level where the two curves intersect is the lysocline. Water is supersaturated with the mineral calcite above the lysocline, undersaturated below it. There is evidence that after the last glacial era the lysocline was deeper than it had been before and than it is now. The carbonate content of deep water must have been greater (*broken line*).

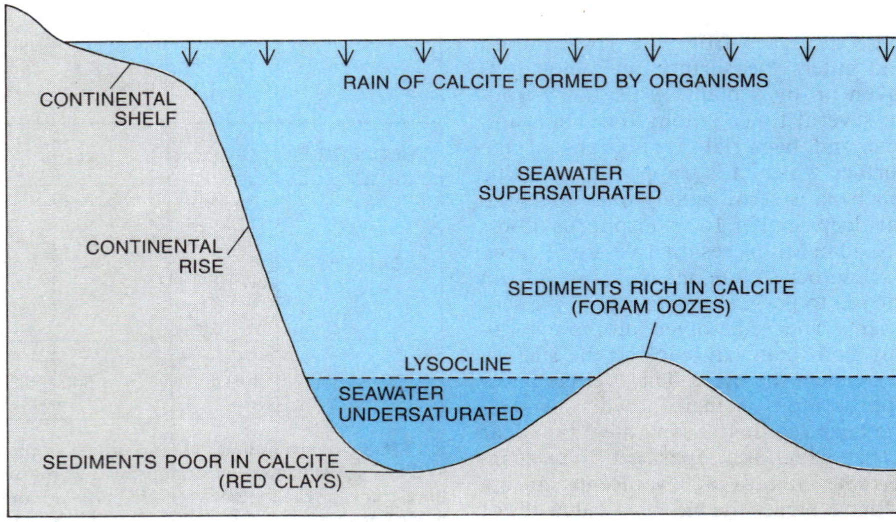

LYSOCLINE separates the ocean's sediments into two broad categories. Sediments below the lysocline, where the water is undersaturated with respect to calcite, have lost their calcite to dissolution, and so they are dominated by clay minerals eroded from the continents: the red clays. Sediments above the lysocline, laid down in water that is supersaturated with respect to calcite, are dominated by the undissolved calcite chambers that were formed by plankton. These sediments are often speckled with the shells of foraminifera, and so they are called foram oozes. The lysocline moves in response to changes in the chemical composition of sea salt, always seeking a level that allows the loss of carbon from ocean water to match the rate of input of carbon to the ocean. In today's ocean marine organisms are producing calcite at a rate that takes up carbon faster than it is supplied by erosion and volcanism. The ocean's chemistry has adjusted itself so that this overproduction of calcite is compensated for by the dissolution of calcite.

Above the lysocline, the depth at which the two curves cross each other, the water is supersaturated with calcite; below it the water is undersaturated. As a result most of the calcite raining onto sea-floor ridges and plateaus above the lysocline is preserved; its carbonate is buried and removed from circulation. Below the lysocline most of the calcite is dissolved and its carbonate is returned to circulation.

Because surface water is supersaturated with calcium carbonate, organisms can manufacture as much calcite as they require. In today's ocean, organisms annually take up several times as much carbon to form calcite as is supplied by erosion and volcanism. The overproduction of calcite is compensated for by changes in the ocean's chemistry that keep the burial of calcium carbonate in balance with the supply of carbon. The lysocline moves up and down in response to the changing concentration of carbonate ion, seeking the level at which the amount of calcite that dissolves just matches the overproduction of calcite by organisms. If, for example, the burial rate of calcite were temporarily to become higher than the rate of supply of carbon destined to be incorporated in calcite, the ocean's carbonate content would begin to decrease. This would lead to a rise in the lysocline and a consequent enlargement of the area of the sea floor over which calcite is dissolved instead of being buried. The imbalance between loss and gain would gradually be restored.

The control systems I have been describing make it possible to calculate the concentrations in the ocean of phosphate and carbonate for any given set of environmental conditions. To quantify the bicarbonate (HCO_3^-) content one turns to charge balance. Discounting a host of minor ions, one can say that

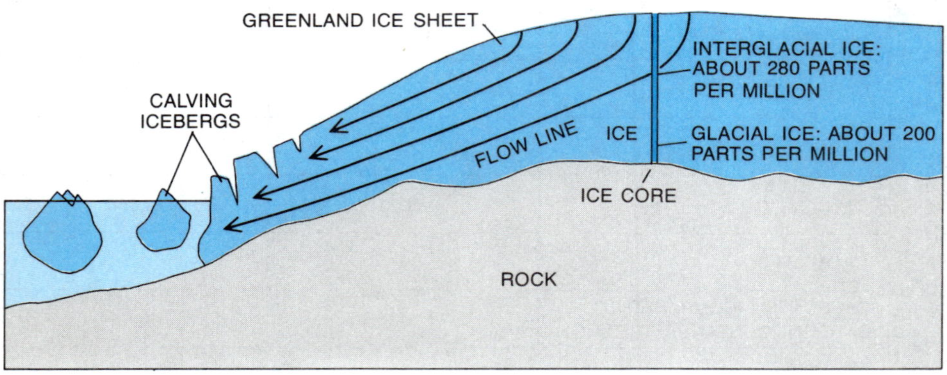

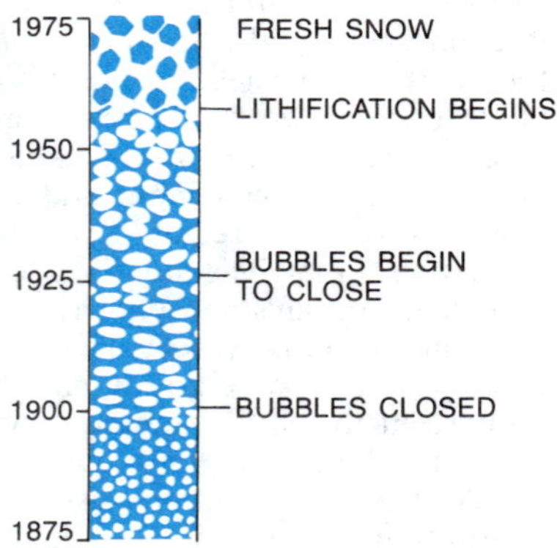

ICE CORES supply evidence for about a 40 percent increase in atmospheric carbon dioxide at the end of the last glacial episode. One deep borehole into the Greenland ice sheet penetrated ice that had accumulated for 80,000 years (*left*). The core contains bubbles of air that were trapped as snow was lithified to become solid ice; such bubbles have been shown to close, and thus seal their air sample, some 50 to 75 years after the snow falls (*right*). In the laboratory of Hans Oeschger of the University of Bern one-centimeter cubes are cut from cores and crushed in a vacuum. The total pressure of the released air is measured and the carbon dioxide content is determined with a laser device. Air trapped in ice some 18,000 years ago, during glacial time, was found to have about 200 parts per million of carbon dioxide; air trapped in the recent past had about 280 parts per million. (Carbon dioxide level has now risen to about 340 parts per million.)

the total concentration of five positively charged ions (sodium, potassium, magnesium, calcium and hydrogen) must equal that of five negatively charged ions (chloride, sulfate, bicarbonate, carbonate and hydroxyl). The hydrogen-ion and hydroxyl-ion concentrations are small enough in seawater to be neglected. On the time scale of glacial episodes on the earth the concentrations of all the other species except carbonate and bicarbonate remain constant. The carbonate ion's concentration has been established. Only the bicarbonate ion remains. Its concentration adjusts to a value corresponding to electrical neutrality, that is, it is equal to the sum of the concentrations of the positive and the negative ions listed above.

At the depth of the lysocline the carbonate-ion content is at the saturation level, and so there the bicarbonate-ion content is fixed by the charge-balance requirement. Having fixed the lysocline-depth composition, one can calculate that of surface water by removing phosphate and the associated carbon dioxide (for soft tissue) and carbonate (for calcite) from the lysocline composition. Now, having fixed the carbonate and bicarbonate content of surface seawater, one has also fixed its carbon dioxide content: since bicarbonate is formed from (and dissociates into) carbon dioxide, carbonate and water, the surface-water concentration of carbon dioxide is equal to the ratio of the square of the bicarbonate-ion concentration to the carbonate-ion concentration (multiplied by a constant that varies with temperature and salinity).

I have not yet mentioned the control system for dissolved oxygen (O_2). The surface water's dissolved-oxygen content is dictated by the solubility of the gas. The deep-water content is set by the phosphate concentration of the deep sea. When deep water reaches the surface, its phosphorus atoms are all converted into (and ultimately lost as) organic residues. The residues are largely consumed by organisms living in the deep sea, creating a demand on the dissolved oxygen there. Overall about half of the oxygen reaching the deep ocean with newly sunken bottom water is consumed by animals and bacteria. The supply depends on the ratio of the amount of oxygen in the air (and hence the amount dissolved in cold surface water) to the amount of phosphorus dissolved in the ocean. Since on the time scale of glacial episodes the oxygen content of the atmosphere cannot change significantly, the dissolved-oxygen content of deep water varies inversely with the amount of dissolved phosphorus.

The dependence of oxygen content on phosphate content reinforces the feedback mechanism tending to balance phosphate loss and gain. In discussing that control mechanism I made the assumption that the fraction of the residues falling into the deep sea and surviving dissolution remains constant even as the water's phosphate content varies. Actually the probability of dissolution is influenced by the degree of oxygenation of the deep water. Since animal life demands oxygen, the efficiency of the scavenging community drops as the oxygen content of the water decreases. Where there is no oxygen there is no animal activity, and the residues are consumed only by the less efficient community of anaerobic bacteria. (In isolated basins where there is no dissolved oxygen at all the fraction of organic residues that are preserved in sediments is observed to increase substantially.) In situations where phosphorus is leaving the ocean faster than it is being supplied by erosion the phosphorus content drops and the deep-water dissolved-oxygen content rises. The decrease in phosphorus reduces the rate of formation of organic residues; the increase in oxygen reduces the fraction of those residues that escape oxidation and are buried. The two feedback mechanisms work in harmony to reduce the rate of loss of phosphorus to sediments, thereby forcing the system back toward balance.

Knowing something about the mechanisms controlling ocean chemistry, one is in a position to evaluate evidence for changes in the past and perhaps to predict changes in the future. What hard

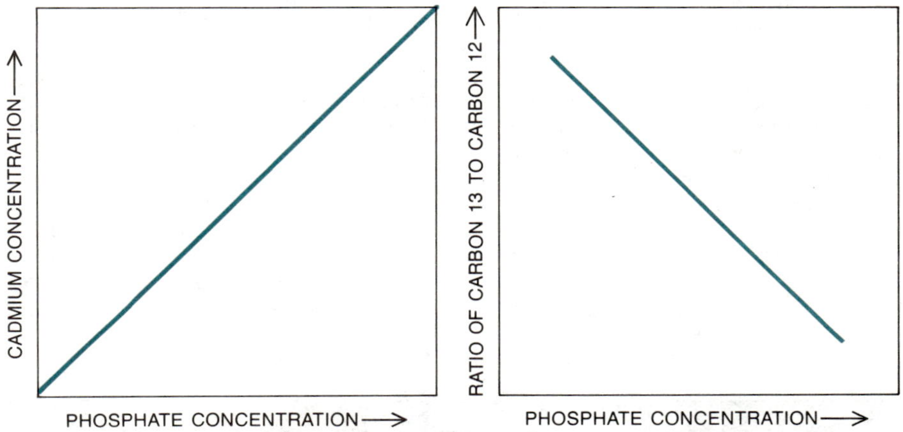

PHOSPHATE CONTENT of the ocean is reflected by two markers in the shells of fossil foraminifera: the content of cadmium and the ratio of the two stable isotopes of carbon. Cadmium is a marker because its concentration varies directly with that of phosphate in seawater (*left*); the ratio of the rarer carbon 13 to the much more abundant carbon 12 in the ocean's total dissolved carbon is a marker because it varies inversely with the concentration of phosphate in seawater (*right*). Analysis of foraminifera from different periods indicates that there was a decrease in the ocean's overall phosphate content at the close of the last glacial period.

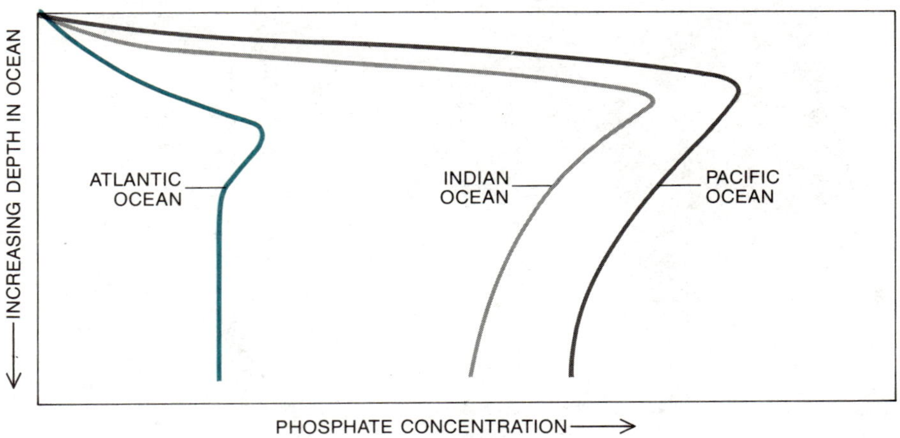

DISTRIBUTION OF PHOSPHATE in the ocean varies with depth. Essentially all the phosphate in surface water is taken up by marine plants and falls into the deep sea in soft-tissue residues. Much of it is released into solution as the residues are decomposed by animals and bacteria, a process that takes place primarily at intermediate depths. Superimposed on this vertical variation there is an Atlantic-to-Pacific increase in the deep-water phosphate content. It results from the flow of newly sunken deep water, much of which now originates in the northern Atlantic and tends to purge the Atlantic of the products of decomposition. The horizontal gradient complicates attempts to reconstruct the chemical composition of the glacial ocean: climatic events may have altered the deep-current pattern as well as the ocean's bulk chemistry.

evidence is there that ocean chemistry has indeed varied over time? The most spectacular and convincing source of information comes from ice cores removed by deep boring into the Greenland and Antarctic ice sheets. The cores reach down to ice that was formed 80,000 years ago. By crushing samples of the ice in a vacuum it is possible to release what is in effect fossil air, trapped in tiny bubbles as new-fallen snow consolidated to become solid ice.

The carbon dioxide content of the released air has been measured, in samples going back 40,000 years, by investigators at the University of Bern and at the University of Grenoble. Both groups got the same striking result: air from ice formed during the peak of the most recent glacial episode, some 18,000 years ago, has only about 70 percent as much carbon dioxide as air from ice formed during the present interglacial period.

An increase of this magnitude in atmospheric carbon dioxide could have been caused by a rise of about 12 degrees Celsius in the temperature of surface ocean water. From rather good independent evidence it is known, however, that the postglacial rise in water temperature was only about two degrees, and the small increase in atmospheric carbon dioxide generated by that amount of warming would have been largely countered by the effect of the oceanwide decrease in salinity resulting from the melting of glaciers. The temperature explanation can apparently be ruled out. The remarkable increase in atmospheric carbon dioxide at the end of glaciation will have to be explained by changes in ocean chemistry.

Another indication of a chemical change on a glacial time scale is a curious large and rather brief deepening of the ocean's lysocline early in the postglacial period. The evidence is in data showing the depth at which the shells of certain marine animals, which would have dissolved had they been below the lysocline, were instead preserved in sediments. The implication is that the concentration of carbonate in the deep sea increased by about 10 or 20 percent at the close of glacial time some 11,000 years ago, reaching a maximum early in postglacial time and then dropping back to about its glacial value perhaps 5,000 years ago. If that is the case, the ocean's calcium carbonate budget must have been seriously perturbed by events associated with deglaciation.

Finally, there are two other items of evidence, which can be interpreted to indicate a substantial decrease in the ocean's phosphate content at the close of the last glacial period. Both come from analysis of the shells of the unicellular marine animals called foraminifera. One finding is a decrease in the content of the element cadmium in the shells of bottom-dwelling foraminifera. The other has to do with the ratio of the two stable isotopes of carbon, carbon 13 and carbon 12, in the shells of foraminifera.

Both cadmium and the carbon-isotope ratio serve as markers for the ocean's phosphate content. The concentrations of dissolved cadmium and phosphate in the sea are closely correlated: both go to zero in the warm surface water, indicating that plants take them up efficiently, and they must have the same pattern of release from organisms deeper in the ocean. Similarly, the ratio of dissolved carbon 13 to carbon 12 is correlated (but inversely) with the phosphate concentration: the higher the phosphate content, the lower the ratio. The reason is clear. In photosynthesis plants take up the abundant lighter isotope, carbon 12, in slight preference to the much less abundant carbon 13. The dissolved carbon of surface waters (where there is almost no phosphate) is therefore slightly enriched in the heavy isotope: its 13-to-12 ratio is higher than the oceanwide average. The organic residues that fall into the deep sea, on the other hand, are eventually oxidized, giving up their excess carbon 12 to the ambient water and thereby lowering the 13-to-12 ratio of the dissolved carbon in

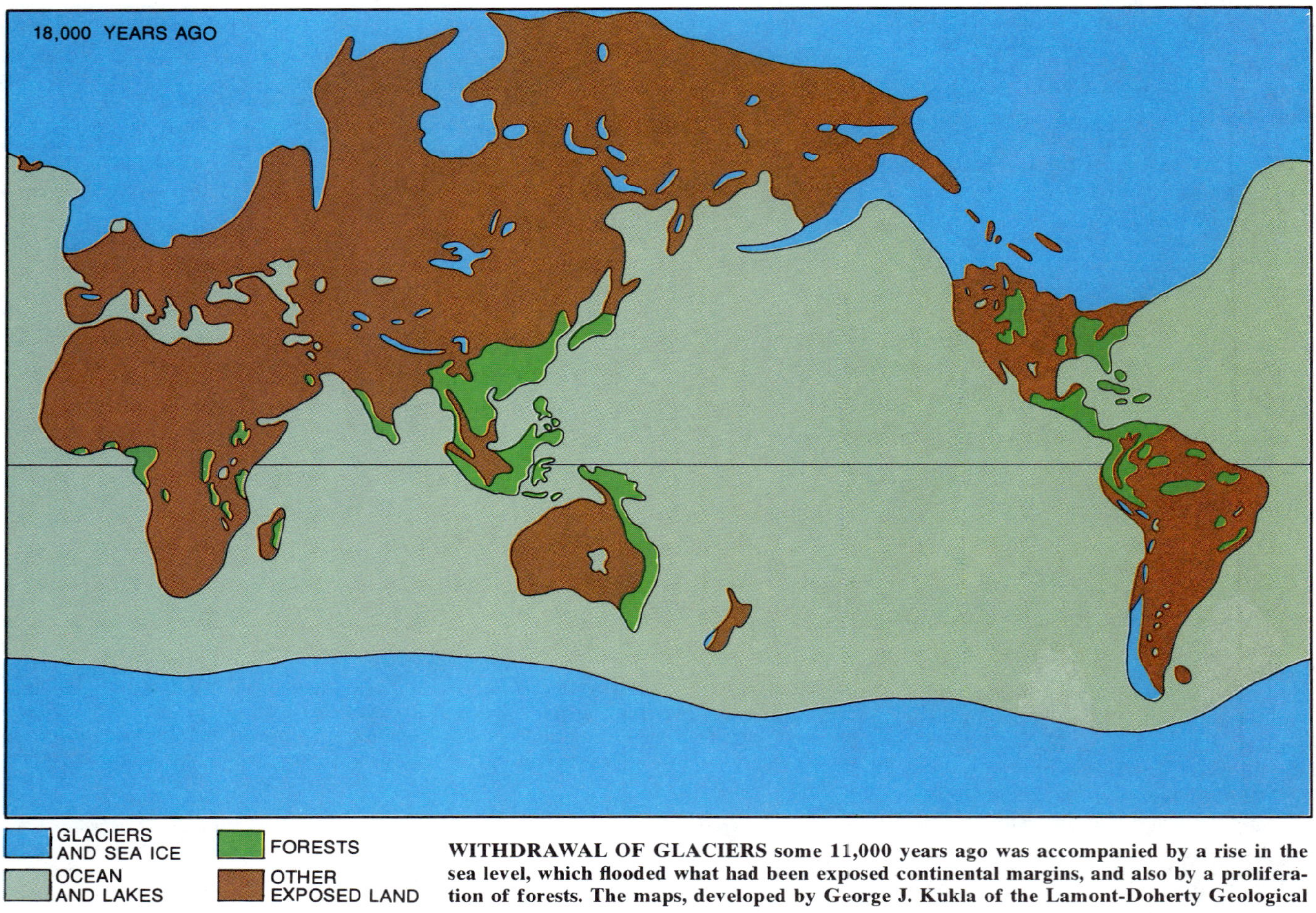

WITHDRAWAL OF GLACIERS some 11,000 years ago was accompanied by a rise in the sea level, which flooded what had been exposed continental margins, and also by a proliferation of forests. The maps, developed by George J. Kukla of the Lamont-Doherty Geological

deep water—where the phosphate content is high.

Edward A. Boyle of the Massachusetts Institute of Technology demonstrated that the cadmium content of foraminifera in recent sediments varies linearly with the cadmium content of the water in which they grew; the content of the element in bottom-dwelling foraminifera shells from deep-sea cores should therefore reflect changes in its concentration in bottom water over time. In cores from various depths in the North Atlantic, Boyle has found a dramatic twofold decrease in foraminiferan cadmium content at the close of glacial time. The implication is that the phosphate content dropped significantly then. How much of the drop resulted from changes in the chemistry of the entire ocean and how much from changes in the pattern of deep-water circulation is not yet known; measurements in cores from other regions should make that clear. At this early stage, however, Boyle's finding provides at least a qualitative indication that there was more phosphorus in the deep sea during glacial time than there was after the end of glaciation.

Examination of sediment cores by Nicholas J. Shackleton of the University of Cambridge has shown that there was an abrupt increase in the ratio of carbon 13 to carbon 12 in the shells of bottom-dwelling foraminifera at the close of glacial time. No such change is observed in the shells of surface-dwelling foraminifera. Because the volume of the deep-water reservoir is much larger than that of the surface reservoir, the mean 13-to-12 ratio of the carbon in the ocean as a whole (and in the atmosphere) must have increased at the close of glacial time. The only reasonable explanation seems to be a net loss of carbon dioxide from the ocean-atmosphere system to organic matter, which would have taken up carbon 12 preferentially; the carbon left in the atmosphere and left dissolved in the ocean would have been slightly enriched in carbon 13. The manufacture of organic carbon could have taken place in the forests that proliferated with the withdrawal of the glaciers. I shall have more to say about this possibility below.

The alternative is that organic carbon manufactured by marine plants in coastal waters was sequestered by burial and storage in shallow sediments, removing carbon—and preferentially carbon 12—from the ocean. Such sediments would have formed on the continental shelves as the glacial ice sheets melted, causing the sea level to rise and transgress the land. The shelf-storage hypothesis, which assumes that the oceanwide change in the carbon-isotope ratio resulted from the storage of organic residues in continental-shelf sediments, can account for all the other chemical changes I have mentioned, provided that cadmium, phosphorus and carbon were present in roughly the same ratios in the trapped residues as they are in residues now falling from the surface into the deep sea.

Even the fact that no postglacial change is observed in the carbon-isotope ratio of surface-dwelling foraminifera can be explained by shelf storage. Because the ratio of phosphorus to carbon in organic residues is about 10 times as high as the ratio in seawater, the sequestering of organic residues removes relatively more phosphorus than carbon from seawater, and so the ratio of phosphorus to carbon would have been reduced by the shelf-storage event. Since phosphorus is the limiting agent for life, the reduction in the ratio would have reduced the extent to which upwelling oceanic carbon was incorporated into organic matter. This in turn would have caused the increase in the carbon dioxide content of surface water, and hence of the atmosphere, that is recorded in ice cores. Another result would have been a decrease in the ratio of carbon 13 to carbon 12 in surface water, which would have canceled the oceanwide increase recorded in bottom-dwelling foraminifera. Still another result would have been

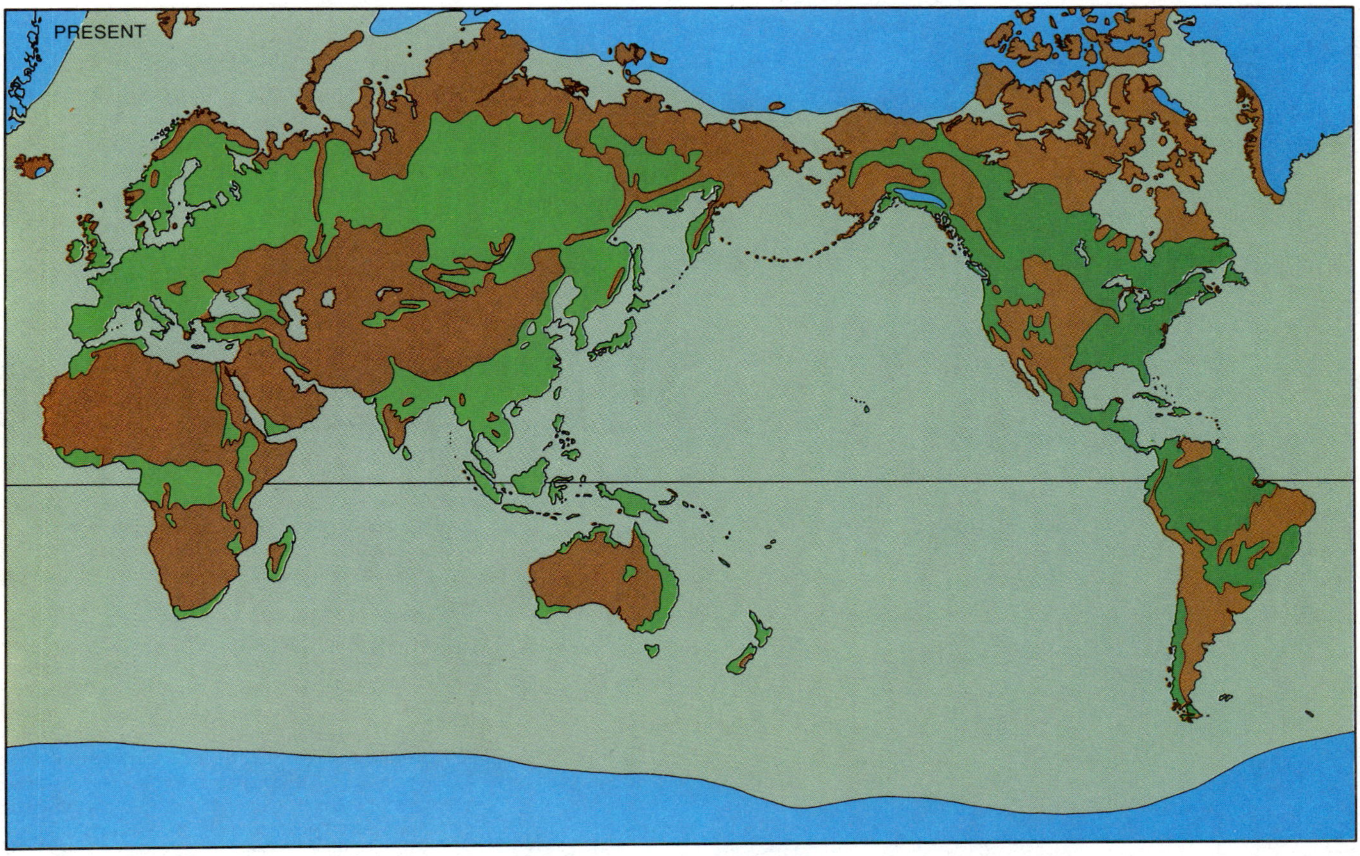

Observatory of Columbia University, contrast conditions at the peak of the glacial era (*left*) with conditions now (*right*). (The current map shows "potential" forestation in the absence of human interference.)

One hypothesis to explain chemical changes at the end of the glacial era depends on the burial of organic matter on newly submerged continental shelves. Another depends on the proliferation of forests.

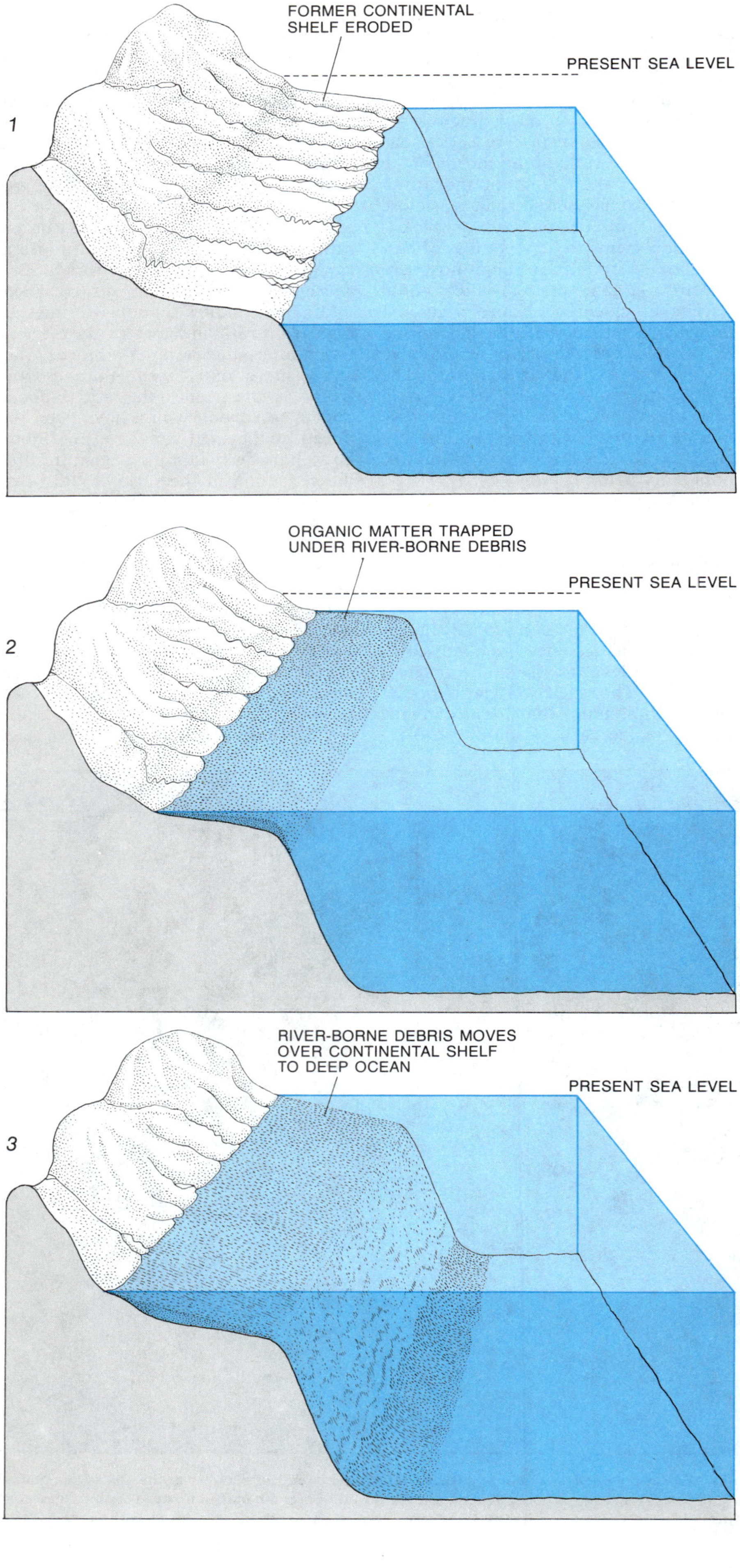

the reduction of the cadmium content of ocean water.

The temporary deepening of the lysocline in early postglacial time could also have resulted from shelf storage. The uptake of carbon by organic matter that was subsequently buried would have left less carbon in the ocean to form the negative ions carbonate and bicarbonate. The charge demand of the ocean's excess positive ions would have forced a larger proportion of the carbon ions to be in the doubly charged state: carbonate rather than bicarbonate. The increase in carbonate would have deepened the isocline. A deeper isocline, however, means more deposition of undissolved calcium carbonate formed by marine organisms, and so the excess carbonate content would gradually have been reduced. In several thousand years the isocline would have returned to about its previous level, just as the fossil record indicates.

I have mentioned the possibility that the expansion of forests at the close of glacial time could have accounted for the reduction of the carbon-isotope ratio. The trouble is that the forests would have taken up from the ocean-atmosphere system only carbon dioxide and not phosphorus. If the growth of new forests had been the only perturbation, the carbon dioxide content of the air would have decreased, whereas the ice cores say it increased.

There is, however, a scenario according to which a combination of forest expansion and ecological shifts in the ocean could generate all the observed postglacial changes. In order for the scenario to work the ratio of phosphorus to carbon in residues falling into the deep sea would have to have been about 30 percent lower during glacial time than it is today. If marine organisms incorporated more carbon in relation to phosphorus in their soft tissue during glacial time and relatively less carbon after glaciation, the ocean's content of carbon dioxide (and therefore the atmosphere's

SHELF-DEPOSITION HYPOTHESIS cites this sequence of events to account for changes in ocean and atmospheric chemistry. During full glacial time (*1*) the sea level was about 150 meters below its present level. What had been (and is now) the continental shelf was exposed and underwent erosion. During the transition to interglacial time (*2*) the sea rose, flooding the sculptured shelf and river estuaries. Organic matter from the coastal ocean was trapped as river-borne debris rebuilt the eroded margin, and the burial of organic matter reduced the carbon and phosphorus content of sea salt. After the sea level stabilized some 6,000 years ago (*3*) the continental margins reached a steady-state shape, allowing river debris to be carried over the shelf into deeper water; the rapid drain of carbon and phosphorus from the ocean came to an end.

carbon dioxide content) would indeed have increased. There is now no way of saying whether or not such a large change in the phosphorus-to-carbon ratio in organisms could have taken place.

Both the shelf-storage hypothesis and the one based on forest growth and the phosphorus-to-carbon ratio can explain most of the observed postglacial changes, and so one cannot yet choose between them. Fortunately the two mechanisms would have had very different effects on the ocean's phosphorus content, and therefore also on its cadmium. The shelf-deposition hypothesis predicts a change in the ocean's cadmium content, whereas the residue-ratio hypothesis predicts no such change. The measurement of cadmium in bottom-dwelling foraminifera from the Indian Ocean and the Pacific, which is now in progress in Boyle's laboratory, should point to one hypothesis or the other.

What is currently being learned about the chemical changes in the ocean at the end of the most recent glaciation may make it possible to generalize somewhat about chemical changes associated with other glacial cycles. It is impossible, however, to draw up a scenario going back beyond the series of major glacial cycles that have dominated the past million years of the earth's history. For one thing the carbon dioxide record in ice now goes back only 40,000 years; even if data could be extracted from older samples, the oldest ice on the earth (deep in the middle of the Antarctic ice sheet) is probably no more than half a million years old. Moreover, the number of variables would keep increasing; for example, neither the concentration of calcium ions in the sea nor the atmosphere's oxygen content could be assumed to remain constant. On even longer time scales, say tens of millions of years, such factors as rates of erosion and rates of deep-sea oxygenation must have been different from what they are now because the basic geometry of the ocean basins and continents was different. New and powerful indicators of paleochemistry will have to be discovered before changes in ocean chemistry can be read reliably over these very long time scales.

This is not to say that isolated, dramatic changes in ocean chemistry may not be detectable if they attended catastrophic events that left clear evidence in ancient fossil records. Luis W. Alvarez, Walter S. Alvarez and their colleagues at the University of California at Berkeley have shown that the profound extinctions marking the close of Cretaceous time 60 million years ago were quite likely triggered by the impact of a large extraterrestrial object, either an asteroid or a comet. The sedimentary record reveals that the event took a particularly large toll on marine life: more than half of the Cretaceous species suddenly disappeared.

If the shock to the web of life was strong enough to have significantly reduced the efficiency with which upwelling phosphorus was incorporated by surviving marine plants, the chemistry of the surface ocean would have changed dramatically, becoming more like that of the deep sea. By analogy with today's ocean one can imagine such a change would have caused a sharp rise in atmospheric carbon dioxide. If the reduction in plant life continued long enough, it would be recorded as a change in the ratio of carbon 13 to carbon 12 in the calcite formed by surviving planktonic organisms. The current worldwide search for evidence bearing on the event that ended the Cretaceous period may turn up such a carbon-isotope anomaly. Even if it does not, it should still yield valuable information about the sequence of events immediately following the impact disaster, and perhaps some clues to chemical changes associated with other such disasters.

In this article I have treated variations in ocean and atmospheric chemistry as the result of chance perturbations in factors controlling the flow of material through the sea to its sediments. J. E.

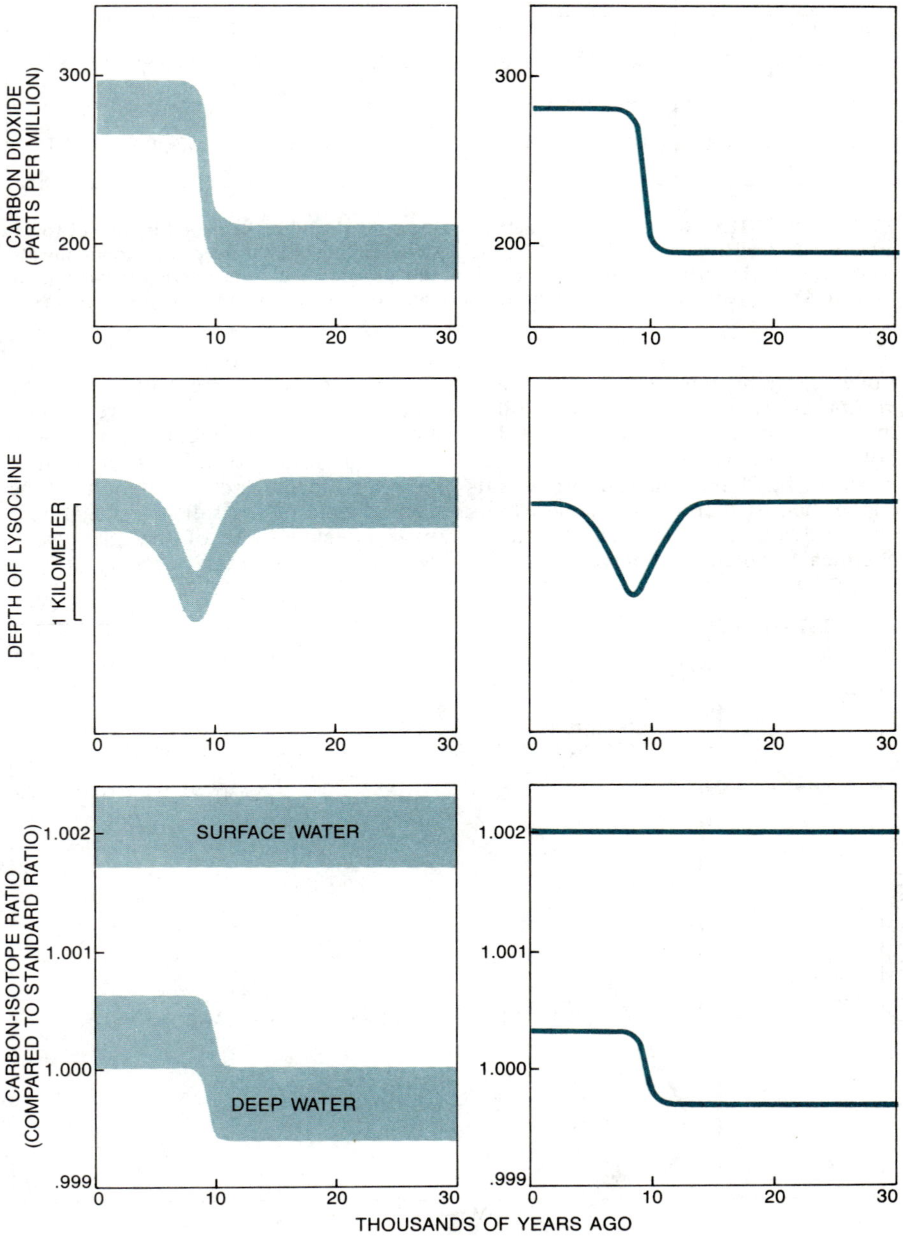

GEOLOGIC RECORD gives observed values (*left*) for the atmospheric content of carbon dioxide (*top*), the depth of the lysocline (*middle*) and the carbon-isotope ratio (*bottom*). The range of observed values brackets the values (*right*) calculated when a two-reservoir ocean is assumed to have been perturbed (by the removal of about 3 percent of the carbon that was present in glacial time and about 30 percent of the phosphorus) as a result of shelf deposition.

	SHELF-DEPOSITION HYPOTHESIS	FOREST AND RESIDUE-RATIO HYPOTHESIS	OBSERVED
PHOSPHATE	DOWN 30 PERCENT	NO CHANGE	NO EVIDENCE
CADMIUM	DOWN 30 PERCENT	NO CHANGE	DOWN 50 PERCENT (IN DEEP ATLANTIC)
CARBON 13 / CARBON 12 (MEAN OCEAN)	UP .07 PERCENT	UP .07 PERCENT	UP ABOUT .06 PERCENT
CARBON 13 / CARBON 12 (SURFACE OCEAN)	NO CHANGE	NO CHANGE	NO CHANGE
LYSOCLINE DEPTH	DOWN 1 KILOMETER	DOWN 1 KILOMETER	DOWN 1-2 KILOMETERS
TOTAL DISSOLVED CARBON	DOWN 3 PERCENT	DOWN 3 PERCENT	NO EVIDENCE
ATMOSPHERIC CARBON DIOXIDE CONTENT	UP 40 PERCENT	UP 40 PERCENT	UP 40 PERCENT

TWO HYPOTHESES, one based on shelf deposition and the other on a combination of forest growth with an altered composition of organic residues falling into the deep sea, explain the observed changes in carbon-isotope ratio, depth of the lysocline and atmospheric carbon dioxide content. They predict different changes in phosphate (and so in cadmium) content, however.

Lovelock, in his book *Gaia: A New Look at Life on Earth,* raises the possibility that the community of living things has not only interacted with the inorganic world but has somehow orchestrated the chemistry of the atmosphere to its own advantage. Perhaps. Certainly chemical change has played a major role in the development of life on the planet. The ocean-atmosphere system has had to adjust its chemistry over rather a wide range to maintain a balance between the loss and gain of many of the constituents of sea salt, and organisms have effected some of the changes as well as being affected by them.

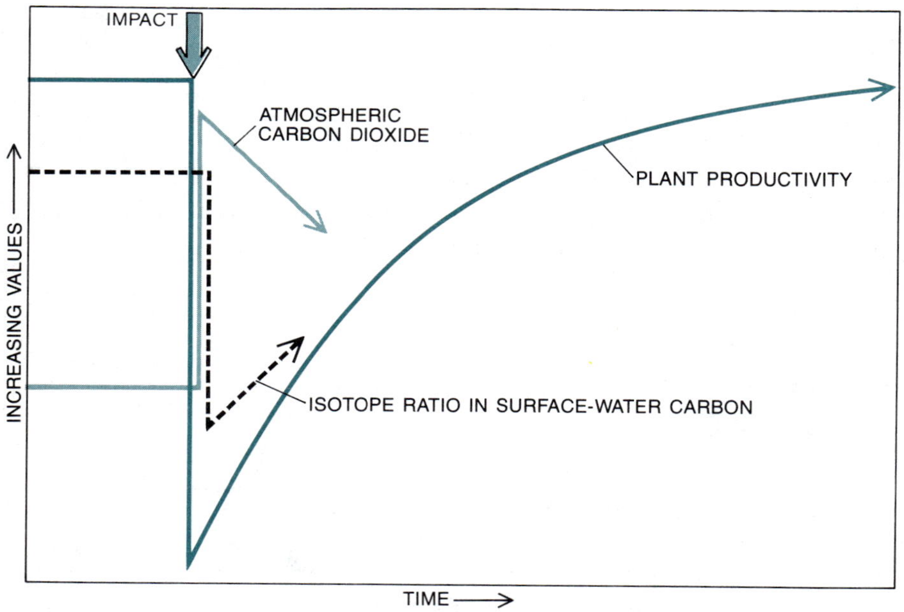

CATASTROPHIC IMPACTS such as the one thought to have ended the Cretaceous period might have generated changes in ocean chemistry that are recorded in the fossil record. If the effects of an impact temporarily devastated marine plant life, there would have been an abrupt increase in surface-water carbon dioxide and in atmospheric carbon dioxide. Eventually plant productivity would have been restored, reducing the carbon dioxide level. Changes in plant productivity should be recorded as changes in the carbon-isotope ratio in foraminiferan shells.

7

THE ATMOSPHERE

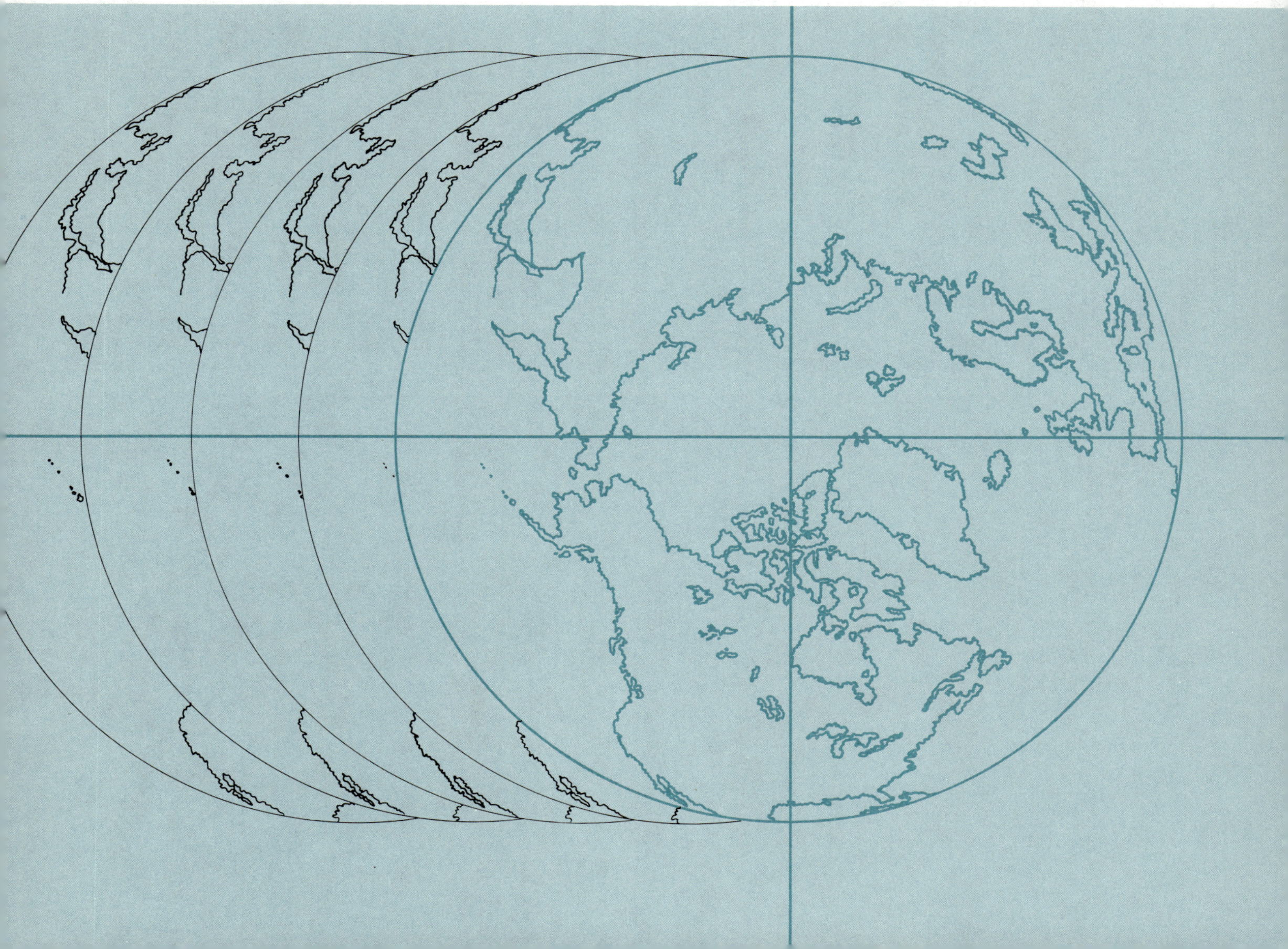

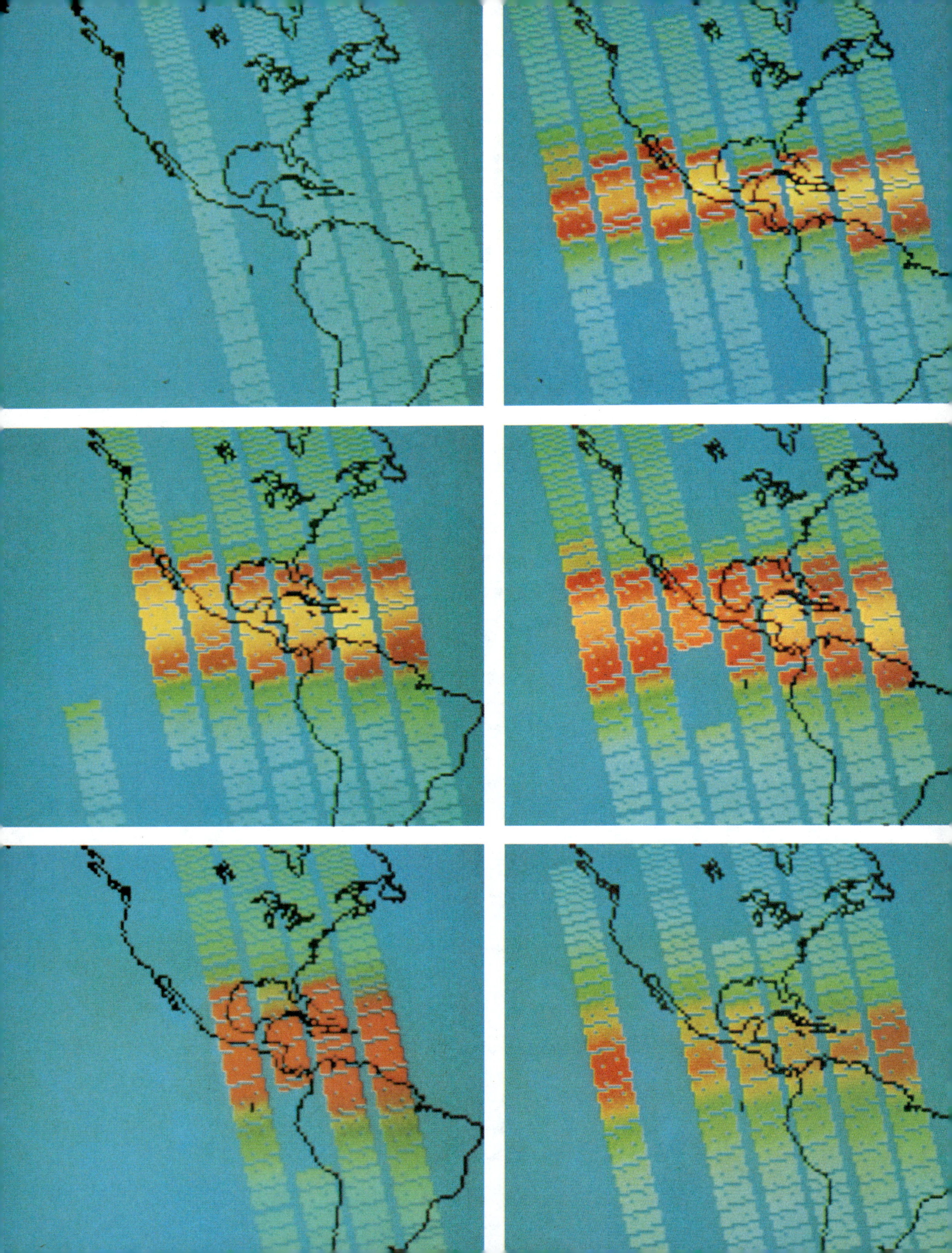

The Atmosphere

by ANDREW P. INGERSOLL

Its dynamic activity serves to distribute the energy of solar radiation received by the earth. Models of this activity help to explain climates of the past and predict those of the future

The atmosphere is the working fluid of the earth's heat engine. Most of the radiant energy arriving from the sun is converted into atmospheric heat energy before it is reradiated into space. The winds redistribute this energy, dissipating more of it in the process than is dissipated by ocean currents, tides, continental drift and mantle convection combined. Both the short-term fluctuations of the atmospheric system (the weather) and the longer-term fluctuations of the average weather (the climate) are an important part of earth history.

The composition of the atmosphere bears little relation to that of the cloud of gas and dust out of which the earth formed. Rather, the composition of the air reflects a complex history of the reactions of volatile elements with the dust in the primordial solar nebula, with magma rising from the earth's mantle, with the rocks of the earth's crust, with the ocean and with the biosphere.

The three major constituents of dry air are nitrogen (N_2), oxygen (O_2) and argon (Ar), which account respectively for 79 percent, 20 percent and 1 percent of the molecules. Nitrogen is for all practical purposes geochemically inert and has therefore accumulated in the atmosphere. Oxygen, on the other hand, cycles through the atmosphere and also through the ocean, the biosphere and sedimentary rocks. The amount of oxygen in the atmosphere is set by the rate of reactions that link the atmospheric reservoir of free oxygen to the reservoir of reduced carbon in sedimentary rocks. The small fraction of organic matter buried in sediments before it can decay corresponds to a net addition of oxygen to the atmosphere; the rate at which oxygen is produced by this means is balanced by the rate at which it is consumed by the weathering of sedimentary rocks. Most of the argon in the atmosphere is the isotope argon 40, produced by the radioactive decay of potassium 40 in the mantle and the crust and released into the atmosphere by volcanoes. Argon, one of the noble gases, is geochemically inert; once it is released into the atmosphere it stays there.

The composition of dry air is remarkably constant over the globe but the amount of water vapor in the atmosphere varies widely, ranging from 4 percent by volume to a small fraction of 1 percent. The atmosphere is not the dominant reservoir of water because water condenses at average earth temperatures. The ocean holds some 300 atmospheres of water (that is, enough water to equal 300 times the mass of all the constituents of the atmosphere), and clays and other hydrous minerals hold a somewhat smaller amount. The water-vapor content of a given air mass is therefore determined by the history of its contact with the surface reservoirs of water and by the saturation vapor pressure of air at different temperatures.

The remaining constituents of air are present in such small amounts that their concentrations are generally given in parts per million rather than in parts per 100 (percent). The most abundant of them is carbon dioxide (CO_2), which at present accounts for about 340 parts per million of dry air. The atmospheric reservoir of carbon dioxide is dominated by the much larger oceanic reservoir of bicarbonate (HCO_3^-) and carbonate (CO_3^{--}) ions and by the reservoir of carbon in limestone rocks, consisting largely of the calcium carbonate ($CaCO_3$) shells of marine organisms. The limestone rocks alone contain some 20 atmospheres of carbon dioxide. The amount of carbon dioxide in the air, like that of water vapor, is therefore set by the rates of the equilibrium reactions linking the atmospheric reservoir to surface reservoirs.

The most abundant gases after carbon dioxide are neon (18 parts per million) and helium (five parts per million). Most of the helium is produced by the decay of radioactive elements in the solid earth. The neon, however, is primordial. The amounts of neon and of the other rare gases (krypton, xenon and the nonradiogenic isotopes of argon) in the atmosphere are probably equal to the amounts of these gases incorporated in the earth at its formation. Once they were released they remained in the atmosphere because they are geochemically inert and do not condense at earth temperatures.

The only other gas with an abundance greater than two parts per million is ozone (O_3). Its abundance varies with altitude, reaching a maximum of 12 parts per million at 30 kilometers. The amount of ozone in the atmosphere is set by the balance between the reactions by which it is created and those by which it is destroyed. The oxygen atoms formed by the photodissociation of mo-

DUST CLOUD INJECTED INTO THE STRATOSPHERE by the eruption of the volcano El Chichón in Mexico on April 4, 1982, was mapped by *Solar Mesosphere Explorer,* a polar-orbiting satellite operated by the University of Colorado at Boulder for the National Aeronautics and Space Administration. The sequence of computer-generated images on the opposite page shows the extent of the cloud at intervals of roughly a month beginning on April 2, immediately before the explosion. The color in the images corresponds to the amount of infrared emission from the cloud, changing from blue to green to red to yellow with increasing radiance. The cloud reached its greatest density in June (*third image*). Thereafter it spread slowly southward and began to disperse. Apparently the pattern of stratospheric circulation kept the dust from moving north of 30 degrees north latitude. The dust particles are highly reflective and decrease the amount of solar radiation reaching the surface of the earth. The net effect of the cloud will therefore be to decrease the mean global temperature (by as much as one degree Celsius). The temperature reductions associated with volcanic eruptions seem to be closely correlated with the sulfur content of volcanic debris. Data from the satellite *Nimbus 7* indicate that El Chichón injected unusually large amounts of sulfur dioxide into the stratosphere. The eruption is therefore expected to cause a relatively large drop in temperature for one of its size.

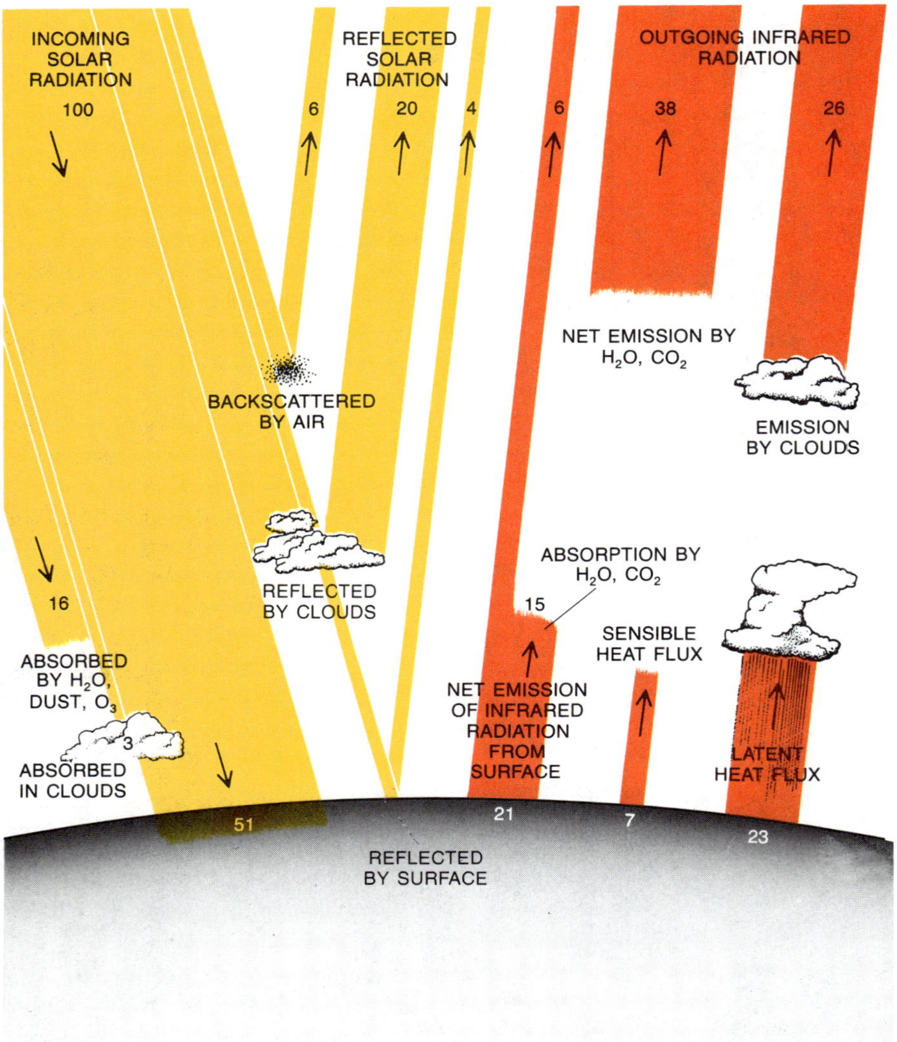

GLOBAL ENERGY BALANCE of the earth and the atmosphere determines the mean global surface temperature and the effective radiating temperature of the earth: the temperature it would appear to have to an observer in space. On the average the earth's surface transfers to the atmosphere an amount of energy equal to the amount of energy it absorbs; the value of the mean global surface temperature, which is about 13 degrees C., is that needed to keep the earth and the atmosphere in thermal equilibrium. On the average the earth as a whole emits into space an amount of radiant energy equal to the amount of radiant energy absorbed by the atmosphere and the earth's surface; the value of the earth's effective temperature, which is about −18 degrees C., is that needed to keep the earth in thermal equilibrium with space. The surface temperature is higher than the effective temperature largely because incident radiation is absorbed at altitudes lower than those from which radiation is emitted into space: the atmosphere is relatively transparent to radiation at visible wavelengths, where the solar emission spectrum peaks, and relatively opaque to radiation at infrared wavelengths, where the radiant emission from the earth is concentrated. (It should be noted that the earth emits about 114 units of infrared radiation; the net infrared emission is the difference between this value and the 93 units absorbed by the atmosphere and emitted back to the surface of the earth.)

lecular oxygen (O_2) by ultraviolet radiation react with molecular oxygen to form ozone. Ozone is destroyed by a number of reactions. Reactions in which "odd oxygen" (either O or O_3) is consumed and molecular oxygen is created lead to a net decrease in the concentration of ozone. Ozone is the only atmospheric gas that absorbs in the near ultraviolet (at wavelengths of from .2 to .3 micrometer) and therefore it plays a crucial role in shielding the surface of the earth from harsh radiation.

The composition of the atmosphere, together with the earth's distance from the sun, determines the atmosphere's energy budget, which in turn determines everything from the temperature of the earth's surface to the pattern of atmospheric circulation that redistributes solar energy over the surface.

The solar flux (the amount of energy carried by sunlight passing perpendicularly through a unit area) at the outer edge of the earth's atmosphere when the earth is at an average distance from the sun in its orbit is 1,367 watts per square meter. It has long been suspected that this "solar constant" is in fact a variable, but only in the past five years has it been possible to measure variations in the constant. Sensitive instruments aboard the *Nimbus 7* and *Solar Maximum Mission* spacecraft have shown that a large sunspot group causes a .1 percent drop in the received radiation. Larger variations may well be revealed by long-term measurements. It is estimated that a 1 percent change in the solar constant sustained for at least 10 years would raise the earth's mean surface temperature by about one or two degrees Celsius.

Not all the radiation reaching the earth is absorbed. About 30 percent of it is reflected back into space by the atmosphere and the earth's surface. The principal reflectors are clouds, atmospheric dust, the molecules of the atmospheric gases, snow and unvegetated land. The percentage of the incident radiation reflected back to space (the planet's albedo) could change substantially if the climate changed, if more dust were injected into the atmosphere by volcanic eruptions or if more land were cleared of forest. A decrease or increase in the albedo would lead to a net heating or cooling of the earth because the sunlight that is not reflected is absorbed.

The earth disposes of the solar energy it absorbs by emitting thermal radiation. Given the solar constant and the earth's present albedo the energy flux averaged over the globe must be about 240 watts per square meter if the earth is to remain in thermal equilibrium. The temperature of the average radiating level can be roughly estimated from the Stefan-Boltzmann law, which states that the flux of radiation given off by a black body (an ideal emitter) is proportional to the fourth power of its temperature in degrees Kelvin. That number (240 watts per square meter) is the power per unit area emitted by a black body at 255 degrees K. (−18 degrees C.). This is the average temperature of the atmosphere at an altitude of five kilometers.

A black body at 255 degrees K. emits radiation at a broad range of wavelengths with a flat maximum at about 12 micrometers (in the infrared region of the spectrum). The earth's surface and atmosphere behave approximately like black bodies and emit radiation in this range. Most of the radiation emitted by the surface is absorbed, mainly by water vapor, clouds, carbon dioxide, dust and ozone. The constituents of the atmosphere that absorb infrared radiation reemit it in all directions. Some of the emitted radiation is absorbed at the surface, adding to the heat energy there, some of it is reabsorbed by the atmosphere and some escapes into space.

If the amount of an infrared absorber

such as carbon dioxide increases, more radiation is absorbed by the earth's surface and its temperature rises. In addition less of the radiation emitted by the surface escapes into space and the earth's stored thermal energy increases; thus infrared absorbers help to warm the earth. The role the atmosphere plays in heating the earth is traditionally called the greenhouse effect, but the term is somewhat misleading. Although the glass of a greenhouse does pass sunlight and block the escape of infrared radiation, most of the heating is due to the fact that the glass roof prevents heat from being dissipated by convection.

The mean global surface temperature (which is currently about 13 degrees C.) is the temperature needed to keep the earth's surface and the atmosphere in thermal equilibrium. The surface, heated by solar radiation and the infrared emitted downward by the atmosphere, transfers on the average an equivalent amount of energy to the atmosphere by evaporation, by conduction, by convection and by the emission of infrared. Local imbalances in the amount of energy absorbed and released by the surface, however, help to create vertical and horizontal temperature gradients in the atmosphere.

The earth absorbs more sunlight at low latitudes than it does at high latitudes. More than half of the sunlight is absorbed by the earth's surface; the rest is absorbed by the atmosphere. The atmosphere at low latitudes and altitudes therefore receives more energy from the earth's surface and from the sun than the atmosphere at high latitudes and altitudes. Moreover, the atmosphere loses more energy to space at high altitudes than it does at low altitudes. As a result of these imbalances temperatures in the atmosphere generally decrease from the Equator to the poles and from low altitudes to high altitudes. The temperature gradients drive the circulation of the atmosphere. The winds generally carry heat down the gradients, from relatively warm areas to relatively cool ones.

The temperature gradients and therefore the large-scale pattern of atmospheric circulation vary with latitude. Within 35 degrees latitude of the Equa-

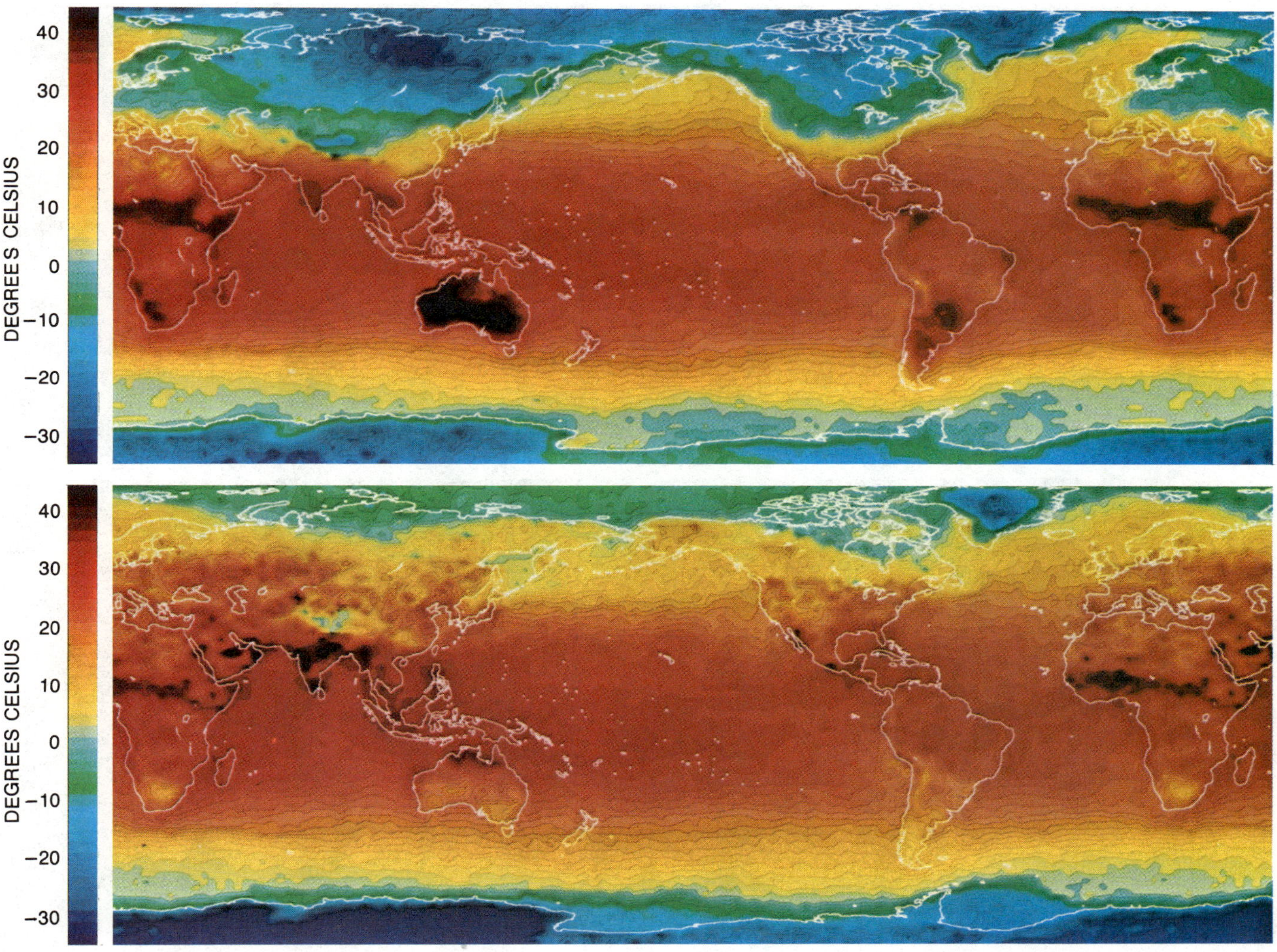

TEMPERATURE GRADIENTS between the Equator and the poles and between the continents and the ocean drive the large-scale motions of the atmosphere. These maps of the mean surface temperature in January (*top*) and May (*bottom*) of 1979 illustrate the gradients, together with the change in them with season. For example, the formation of hot regions in northern India in May is a precursor of the southwesterly winds that blow from the cooler Arabian Sea across much of the Indian peninsula in the monsoon season from June to October. The increase in sea-surface temperatures from winter to summer is much less than the increase of land-surface temperatures largely because wave motion distributes heat to greater depths in the sea. These images were generated from data collected by the infrared and microwave sounding units carried by National Oceanic and Atmospheric Administration weather satellites. The surface temperature was derived from the microwave data by means of an analytic method that compares the data from the microwave sounder with those from the infrared sounder in order to distinguish among the contributions to the measured radiances made by the clouds, the atmosphere and the surface. The method was originally developed by Moustafa T. Chahine of the Jet Propulsion Laboratory of the California Institute of Technology for the purpose of analyzing atmospheric data acquired by space missions to other planets. It has since been applied to satellite data on the earth's atmosphere by Milton Halem and Joel Susskind of NASA's Goddard Space Flight Center.

tor the large-scale winds are meridional (north to south). Two huge circular cells connect the hot moist air rising over the Equator to the warm dry air sinking at higher latitudes. The hot air rising in what is known as the Intertropical Convergence Zone (ITCZ), within 10 degrees north and south of the Equator, cools as it rises, which results in intense precipitation. The ITCZ coincides with the tropical rain forests and regions of heavy precipitation over the oceans. The dry air spreads poleward at high altitudes and sinks throughout the subtropics, the regions between 10 and 35 degrees latitude in both hemispheres. When it reaches the surface again, it is still dry; the great deserts of the world are in these zones.

Poleward of 35 degrees latitude the circulation is predominantly zonal rather than meridional. The flow of air is dominated by the meandering high-altitude jet stream that blows generally eastward in both hemispheres at speeds of up to 45 meters per second (100 miles per hour). Long waves or undulations that begin as small fluctuations and deepen with time are superimposed on the general west-to-east flow. The troughs of the waves are where the jet stream is closest to the Equator; the ridges are where it is closest to the poles. As the waves develop, masses of cool air move toward the Equator behind (to the west of) the troughs and masses of warm air move toward the poles ahead of (to the east of) the troughs. The development of the waves thus contributes to the transport of warm air toward the poles and of cold air toward the Equator. The masses of air trapped in the troughs and ridges are carried eastward by the movement of the high-altitude waves. The waves thereby help to determine the movement of the surface high- and low-pressure regions that dominate the weather over North America and other mid-latitude regions.

Beginning about 50 years ago the dynamics of large-scale atmospheric motions (as they are now understood) began to be formalized in mathematical equations. Among the pioneers in this work were Carl-Gustav Rossby, the Swedish meteorologist for whom long waves in the jet stream are named, and

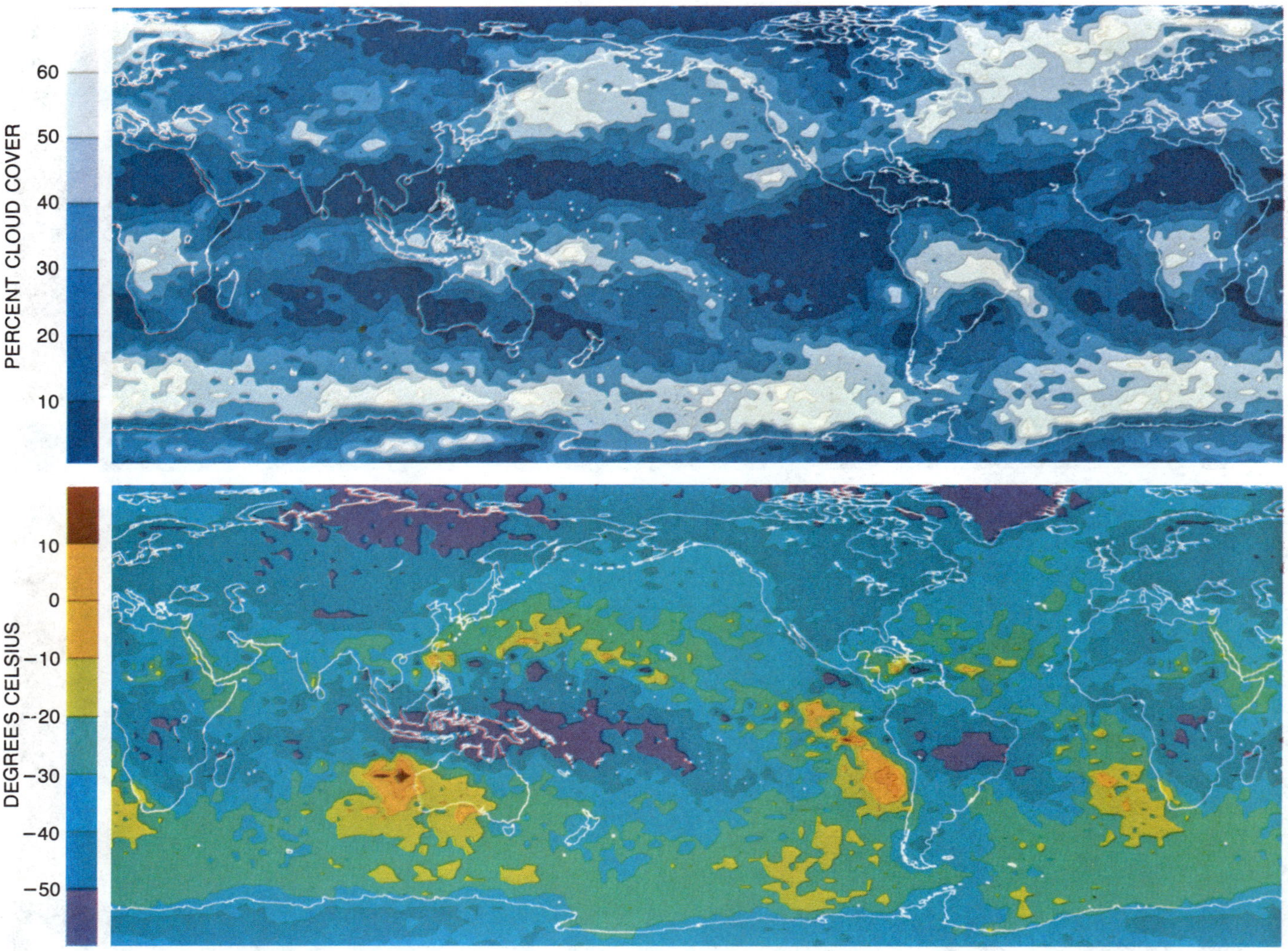

MAP OF MEAN CLOUD COVER in January, 1979 (*top*), generated by the image-processing group at the Jet Propulsion Laboratory from data collected by weather satellites provides an index to characteristic atmospheric circulation patterns and the climate zones they create. For example, the belt of clouds near the Equator is created by the convergence of the updrafts of two huge north-south circulation cells called Hadley cells. The hot air rising in this zone, called the Intertropical Convergence Zone (ITCZ), cools as it rises. As the map of the mean temperature at the top of the clouds in January, 1979 (*bottom*), shows, the cloud tops over the ITCZ are high and therefore cold. The tropical rain forests and comparable regions of heavy precipitation over the oceans are in this zone. The annual precipitation can reach two meters and exceeds the annual evaporation by more than 50 centimeters. The relatively cloud-free strips above and below the ITCZ extending up to about 35 degrees latitude are the regions over which the still-warm but by now dry air in the Hadley cells sinks. The great deserts of the world are in these zones, which are called the subtropical arid zones. The large-scale circulation pattern at higher latitudes is governed by the development of waves in the meandering high-altitude jet streams and is therefore more variable than the pattern at low latitudes. The pattern of waves in the zonal flow at middle latitudes is suggested by such features of the maps as the deflection of the cloud belts (storm tracks) southward along the west coast of North America and northward along the east coast.

Jule G. Charney, who was for many years at the Massachusetts Institute of Technology. Rossby and Charney led the way to the development of workable computer models of the large-scale circulation by developing equations that filtered out the effect of small-scale motions. Earlier atmospheric models had been subject to the explosive growth of errors in part because no such discrimination was made. Today global atmospheric models running on high-speed computers provide the basis for regional and local daily weather forecasts. The limit on useful forecasts is now about a few days to a week and is proving difficult to extend.

The first step in developing a forecast is to summarize the values of different atmospheric variables, such as wind speed and pressure measured at a large number of scattered locations, as the average values that apply at the intersections of an imaginary three-dimensional grid wrapped around the globe. The mathematical model, which is based on simple physical principles such as Newton's second law of motion and the gas law, is then used to determine how conditions at each point on the grid will change in a short interval of time, such as 10 minutes, during which the rate of change can be assumed to be constant. The new values are then used in place of the original ones and the process is repeated through the forecast period.

One of the major problems weather forecasters face, the problem presented by small-scale atmospheric phenomena, may suggest why it is difficult to extend the period of useful forecasts. Small-scale phenomena such as turbulent eddies and convective storms are relatively energetic, as Herbert Riehl of the National Center for Atmospheric Research and Joanne Starr Malkus of the Woods Hole Oceanographic Institution pointed out in the 1950's. For instance, most of the vertical motion in the Tropics occurs in isolated storms covering .1 percent of the area. The latent heat of evaporation released when water condenses in the storm clouds is a major source of atmospheric energy. In addition the clouds reflect sunlight and absorb infrared radiation, thus further altering the distribution of energy that eventually determines the weather. Collectively these small-scale phenomena have a significant effect on global weather patterns, although individually they are below the level of resolution of the weather-forecasting models.

The small-scale phenomena are not ignored in the models; they are simply inadequately treated. The equations that are applied to tracking large-scale motions in the atmosphere are deterministic; the values of the parameters in the grid at any one time determine those in the grid after the next cycle of calculations. Small-scale processes, however, are handled by statistical subroutines of the models that specify the most probable net effect of all small-scale processes inside a cube on the basis of the values at the cube's corners. Sometimes the most probable events do not occur, and statistical errors therefore contribute significantly to the rate at which the evolution of the model diverges from that of the atmosphere.

It is unlikely that this problem will be solved simply by increasing the level of resolution of the atmospheric models. The resolution of a model, which is determined essentially by the size of a cube in the grid, is ultimately limited by the number of arithmetic operations needed to keep track of the evolution of the variables in the grid. For example, keeping track of seven atmospheric variables (temperature, pressure, water vapor, cloud cover and the wind speed along three axes) in a grid consisting of cubes 200 kilometers square and 10 layers deep actually means keeping track of a million variables. Interactions with nearby variables largely determine how each variable changes with time. Some 500 arithmetic operations are required to compute the interactions to which one variable is subject, so that a total of about .5 billion operations must be done for each 10-minute time step of the model. A tenfold increase in the spatial resolution of the three-dimensional grid would increase the number of variables by three factors of 10, and the number of time steps per hour would have to increase by a comparable factor. The number of arithmetic operations would therefore increase 10,000-fold, and yet such a model would still not resolve atmospheric phenomena smaller than 20 kilometers.

The problem of small-scale atmospheric phenomena and other problems that limit the period of useful forecasts

GLOBAL ATMOSPHERIC MODEL developed by the European Centre for Medium Range Weather Forecasts, an organization supported by 17 European nations, produced the data from which this image of predicted global cloud cover was generated. Measurements of atmospheric variables such as temperature, pressure, wind speed and moisture content, made at many locations scattered around the world, are summarized as the average values that would hold at the intersections of a three-dimensional grid wrapped around the globe. The model calculates future conditions from the initial conditions by means of equations ultimately based on simple physical principles such as Newton's laws of motion, the thermodynamic equations and the gas law. The resolution of the image suggests the limit to the resolution of global forecasting models imposed by the enormous number of calculations needed to keep track of the evolution of the data points. For example, the European Centre's Cray-1 computer must do 500 billion arithmetic operations to predict the behavior of the atmosphere 10 days later.

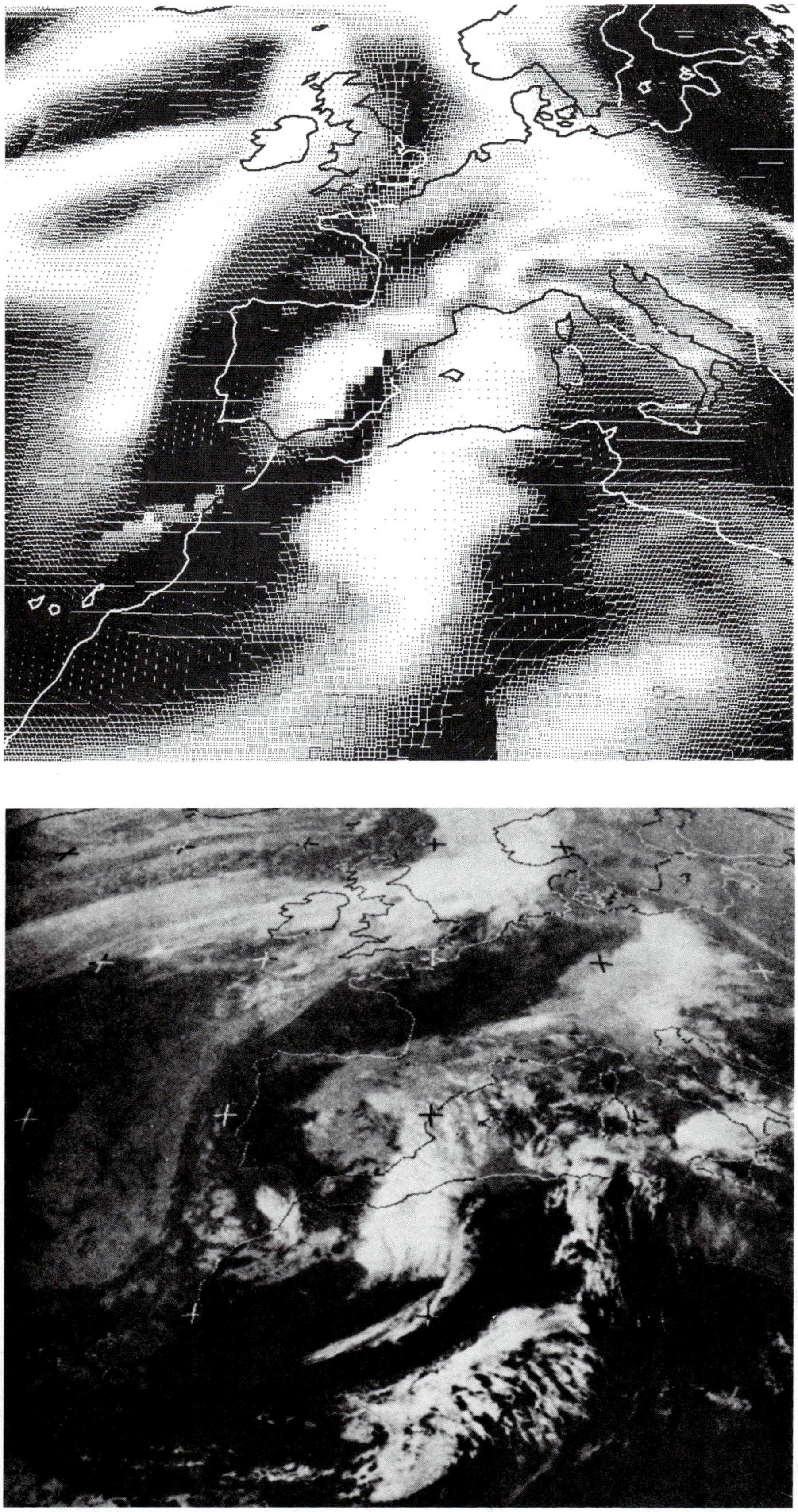

HIGH-RESOLUTION ONE-DAY FORECAST of cloud cover produced by a limited-area version of the European Centre for Medium Range Weather Forecasts' atmospheric model (*top*) compares well with an image of cloud cover made by the satellite *Metosat* one day later (*bottom*). The grid used for limited-area forecasts consists of cubes 50 kilometers square (less than .5 degree in latitude or longitude) and is 15 layers deep. The cloud system approaching Europe from the Atlantic is a cold front: advancing cold air is underriding retreating warm air.

are being addressed by a number of major research programs. The experiments being conducted as part of the Global Atmospheric Research Program (GARP), organized by the World Meteorological Organization and the International Council of Scientific Unions in the early 1960's, are basically designed to provide data about atmospheric phenomena that will allow better forecasting models to be developed. The GARP Global Weather Experiment conducted in 1979, for example, consisted of two two-week intensive observation periods. During these periods atmospheric variables were measured by balloons, satellites, aircraft, ships and buoys. Parameters of the sea, land, ice, soil and vegetation that might have an impact on the weather were measured as well.

One goal of the experiment was to see how sensitive the forecasts for one part of the globe, say North America, are to initial conditions over other parts of the globe, say the normally ill-defined conditions over the South Pacific. Another was to determine whether measurements of variables, such as soil moisture, that are not now exploited in weather forecasting would improve forecasting accuracy. A third goal was to provide more detailed data on the troublesome small-scale atmospheric phenomena.

Among the research programs in the U.S. directed at improving global forecasting models are those at the National Center for Atmospheric Research in Boulder, Colo., the Goddard Space Flight Center of the National Aeronautics and Space Administration and the Geophysical Fluid Dynamics Laboratory of Princeton University. One of the problems workers at these institutions are addresssing is that of determining which of a multitude of possible sources of error is actually responsible for erroneous forecasts.

Is it the inaccuracy of the initial observations, the limitations of the models or the unpredictability of the atmosphere itself that is currently setting the upper limit on the period of useful forecasts? There is no simple answer to the question, but a test applied by Edward N. Lorenz of the Massachusetts Institute of Technology, who was one of the first to make estimates of this kind, may suggest what can be done to pin down the sources of inaccuracy. If computations are run on two sets of initial weather conditions that differ by an amount within the estimated uncertainty of the observations, the rate at which the calculated conditions then diverge yields a measure of the forecasting inaccuracy due primarily to observational error. Current atmospheric models double small differences in initial conditions every two to three days. Assuming that the observational errors are distributed uniformly over the globe and that the evolution of models exactly duplicates the evolution of the weather, this doubling rate suggests one-to-two-week forecasts (several doubling times) are theoretically possible.

How can forecasting models be improved? Given the complexity of the system the models seek to simulate, the best approach may be not more exhaustive calculations but rather calculations discriminating more accurately between phenomena that have a significant impact on the development of the weather and phenomena that have an insignificant impact, in effect improved recognition of patterns. Patterns that persist for more than a few days, called blocking phenomena, might include waves in the jet stream, the paths that storms habitually take or areas of persistent drought.

Conditions in the upper levels of the ocean may also provide a basis for long-range weather forecasting. Anomalies in sea-surface temperatures and in the slope of the sea caused by the prevailing winds can persist for months or years. Anomalies in the surface temperature could be exploited to gauge the amount of energy that will be supplied to the atmosphere as the latent heat of evaporation. More generally, such anomalies can trigger chains of events that unfold in a predictable way, such as the sudden change in ocean currents and weather patterns near the coastline of Peru called an El Niño event. The major El Niño event of 1983 is providing an opportunity to study the causes and possible consequences for global weather patterns of such changes in ocean temperature and circulation patterns.

Although the weather is highly variable, large deviations in temperature and precipitation are usually followed by a return to average or normal values. In other words, the long-term weather or climate seems relatively stable. The geologic evidence, however, indicates that the earth's climate does change radically on the scale of thousands and millions of years.

Perhaps the most dramatic example of such a change is the recent ice ages. For most of its history the earth seems to have been largely ice-free. Beginning several million years ago, however, it started to undergo cyclic glaciations. Over periods ranging from 20,000 to 100,000 years ice sheets build up on the land masses of the Northern Hemisphere. (Southern Hemisphere glaciers grow at the same time, but they account for much less of the global ice volume because there are no large land masses from 40 to 70 degrees south latitude on which glaciers can grow.) In the Northern Hemisphere the ice builds up to a thickness of three kilometers, and the weight of the ice depresses the crust under it by as much as 500 meters. When the ice becomes thick enough, the pressure of its weight causes it to flow outward and southward until it covers large areas of North America and Eurasia. The sequestering of water in the ice lowers the level of the oceans by 100 meters or more. At the height of a glacial period the average temperature of the earth is two or three degrees C. lower than normal. The ice then retreats abruptly. The length of the retreat phase is uncertain, but the geologic evidence suggests it is several thousand years or less.

What causes the climate to go on such a rampage? The answer seems to lie in the response of the atmosphere to external forcing mechanisms such as changes in the amount or distribution of solar radiation. The external changes by themselves are too small to account for the large changes in climate; what appears to matter more are how these

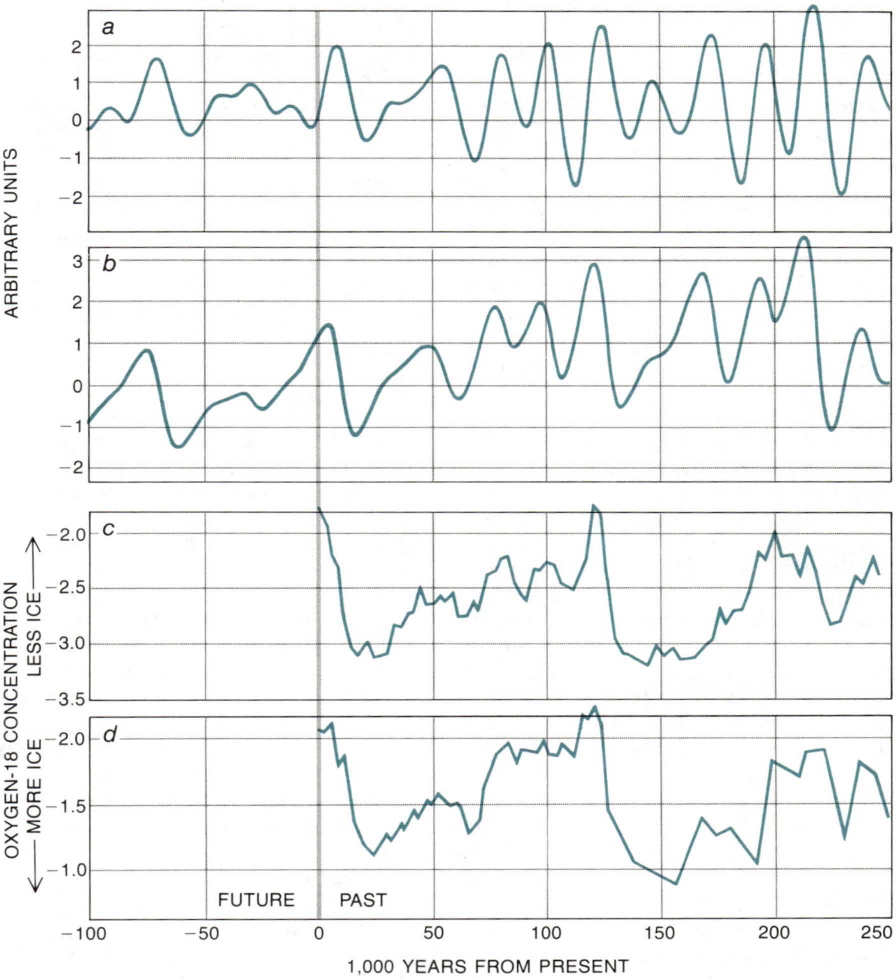

MILANKOVITCH THEORY of the climate cycles during the recent ice age holds that the forcing mechanism is the change in the amount of Northern Hemisphere summertime insolation (incident solar radiation) caused by the variation of three parameters of the earth's orbit around the sun. The two parameters that determine the period of fluctuations in Northern Hemisphere insolation are the tilt angle of the axis, or the angle between the axis and the plane of the ecliptic (which determines to what extent the North Pole is pointed toward the sun in the Northern Hemisphere summer), and the direction in which the earth's axis points (which determines whether perihelion, the time of the year when the earth is closest to the sun, comes in the Northern Hemisphere summer or the Southern Hemisphere summer). The orbital input (*a*) fluctuates with these parameters, which have periods of 40,000 and 20,000 years. The ratio of the heavy isotope oxygen 18 to the lighter isotope oxygen 16 in ocean sediments provides a measure of the global ice volume. As saturated air moves poleward the water molecules incorporating the heavier isotope are preferentially removed by precipitation, so that snow falling at high latitudes is enriched in oxygen 16. The oceans are therefore enriched in oxygen 18 in cold periods when snow accumulates on the land. The isotope ratio in two representative deep-sea cores, one from the southern Indian Ocean (*c*) and the other from the Pacific (*d*), measured by John Imbrie and his son John Z. Imbrie of Brown University, does fluctuate with periods of 40,000 and 20,000 years but would be better characterized as a sawtooth curve with a period of about 100,000 years. A climate model developed by the Imbries (*b*), which assumes a long lag time before ice accumulates in response to the decrease in insolation and a much shorter lag time before the ice melts in response to an increase in insolation, suggests how the changes in incident solar radiation might be translated into the changes in global ice volume.

changes are amplified or damped by the atmosphere, the ocean and the ice. Among the first to attempt to assess the effects that might increase or decrease the sensitivity of climate to external forcing were Mikhail I. Budyko of the Main Geophysical Observatory in Leningrad and William D. Sellers of the University of Arizona. In the 1960's they independently published climate models incorporating feedback effects that governed the climate changes triggered by changes in the solar constant. For example, lower polar temperatures would cause snow and ice to accumulate at high and middle latitudes. The albedo of snow is higher than that of land or water, and so less sunlight would be absorbed at these latitudes, causing the temperatures to drop further.

Formulating models of climate variability remains an uncertain business. For one thing the relative importance of positive and negative feedback effects is difficult to assess. For example, ice-albedo feedback and other factors accentuating the decrease in temperature at high latitudes might be offset by atmospheric and oceanic circulation driven by the gradient of temperature between the Equator and the poles. Climate models are also subject to many of the problems encountered in weather models, problems exacerbated by the relative lack of data on conditions in glacial epochs. In addition climate models must take account of the interaction of the atmosphere with the ocean and the ice sheets, interactions that can be ignored with relative impunity in short-term weather models. Little is known, however, about the dynamics of the deep ocean and the continental ice sheets.

In spite of these uncertainties, comparing climate models with the geologic record of climate change is instructive. The most successful models of the recent ice ages are based on the astronomical theory of climate forcing. This theory concerns changes in the amount of solar energy received by the earth that are associated with three cyclically changing parameters of the earth's orbit around the sun. The first parameter is the tilt of the earth's axis (the angle between the axis and the plane of the orbit), which has a period of about 40,000 years. The second is the direction in which the earth's axis points, which has a period of about 20,000 years. The third is the eccentricity of the earth's orbit (the degree to which it departs from a circle), which has a period of 100,000 years. The orbital forcing can be calculated for millions of years into both the past and the future.

The astronomical theory is also known as the Milankovitch theory, after Milutin Milankovitch, the Yugoslav scientist who made such calculations in the 1920's and 1930's. According to Milankovitch, the key factor for the earth's climate is not the total amount of sunlight received by the globe in the course of the year, which in any case changes only slightly with changes in the orbital parameters, but rather the amount of sunlight received at high latitudes in the Northern Hemisphere during the summer. The tilt angle, which determines to what extent the North Pole is pointed toward the sun during the Northern Hemisphere summer, has the greatest impact on the amount of sunlight received in the summertime. The direction in which the axis points determines whether perihelion, the time of year when the earth is closest to the sun, comes during the Northern Hemisphere summer or the Southern Hemisphere summer. The eccentricity of the orbit affects the amount of sunlight received at perihelion and thus determines the amplitude of the 20,000-year forcing cycle. The eccentricity of the orbit by itself has very little effect on summertime insolation and so there is little climate forcing at the period of 100,000 years.

The clearest corroborating evidence for the orbital forcing of the glacial cycles comes from the ratio in ocean sediments of two isotopes of oxygen: oxygen 18 and oxygen 16. As an air mass moves away from the Equator water molecules incorporating oxygen 18 are preferentially removed by precipitation. The snow that falls at high latitudes holds more of the lighter isotope. Therefore when snow accumulates on the land, the oceans are slightly enriched in oxygen 18. Changes in the oxygen-isotope ratios in deep-sea sediments thus reflect changes in global ice volume. One of the first to recognize the importance of the sedimentary isotope ratios as an index to past changes in the climate was Cesare Emiliani of the University of Miami. The results of systematic oxygen-isotope analyses were published in 1976 by J. D. Hays of Columbia University, John Imbrie of Brown University and Nicholas J. Shackleton of the University of Cambridge.

The isotope record of the past 500,000 years shows prominent oscillations in the global ice volume at periods of 40,000 and 20,000 years. The dominant feature, however, is a steady increase in the ice volume for 100,000 years or so followed by a sudden decrease. Climate models in which the ice simply responds passively to the orbital forcing do not duplicate the isotope record; the ice volume predicted by the models oscillates with periods of 40,000 and 20,000 years in direct response to the forcing. Some other factor or factors must be introduced to account for both the 100,000-year cycle and the abrupt recovery from glaciation.

Models recently published by David Pollard of Oregon State University illustrate the problem and a possible solution. He finds that the earth's climate at present is such that the "normal" state of the earth is icebound. There should be extensive permanent ice sheets. In his "passive" model the global ice volume fluctuates slightly with periods of 40,000 and 20,000 years, but it does not undergo the dramatic losses at periods of 100,000 years recorded in the isotope record. In order to get a more realistic response Pollard introduced a mechanism of rapid ice loss into the model: the calving of icebergs into the ocean causes the ice sheet to become unstable when it reaches a critical size.

The earth's crust does not respond immediately to glacial loading. During extended periods of glaciation, however, the weight of the ice depresses the crust until it is below sea level, although the top of the ice is still much above sea level. At this point the ice is particularly vulnerable to small changes in incident sunlight, such as those that might be caused by orbital forcing. If the ice sheet thins in response to such changes, the crust does not rebound immediately. Along the outer edge of the ice 90 percent of its thickness may be below sea level. Since ice is lighter than water, if the sea intrudes into the crustal depression, the ice will float. Floating ice is assumed to suffer rapid loss along its outer edge as icebergs calve off the ice sheet into the shallow sea. The ice sheet thins further as the thicker ice at the center of the sheet flows outward to establish a new equilibrium. Rapid loss and the consequent retreat of the ice continue until the edge of the ice sheet reaches high ground.

There is evidence that the crust in areas such as Scandinavia and Canada from which the last glacier retreated some 10,000 years ago is still rebounding. This puts the crustal response time into the same range as the one assumed in the model. Still, whether the instability of the ice sheet brought on by the depression of the crust is what causes the retreat of the ice during interglacial epochs remains the subject of speculation.

What is known about the climate of the far distant past, before the ice ages? The climate models devised by Budyko and Sellers predict that the earth would turn into a bright, ice-covered planet if the solar output were a few percent lower. The feedback effects the models incorporate not only increase the earth's sensitivity to the decrease in solar output but also, once the earth cools off, decrease its sensitivity to subsequent increases in solar output. In short, if the solar constant were lower, the earth would be locked in a deep freeze. Has anything of the kind ever happened?

Astronomers know something about the sun's past because they can observe stars that are much younger than the sun but have the same mass and composition. They conclude that the solar output was 40 percent lower shortly after the earth formed some 4.6 billion years ago than it is today and that it has been increasing ever since. Theoretical models of the sun's internal structure and evolution support this conclusion.

Geologists have compiled only a fragmentary record of the climates in the distant past, but the general trend of the evidence is clearly opposite to the one predicted by the trend of the solar constant and models of the earth's response to the forcing mechanism. Samuel Epstein and his colleagues at the California Institute of Technology have inferred temperatures from the oxygen-isotope ratios and comparable ratios of hydrogen 2 (deuterium) to hydrogen 1 (ordinary hydrogen) in cherts (crystalline siliceous sedimentary rocks). From the isotope ratios in a chert 3.5 billion years old they inferred that the temperature of the ocean in which the rock formed was higher than 50 degrees C. Different isotope ratios in more recent cherts suggest that temperatures became progressively lower. From the oxygen-isotope ratios in the sediments formed by the calcium carbonate shells of benthic foraminifera (minute animals of the deep ocean) it was inferred that the temperature at the bottom of the ocean was about 15 degrees C. in the period from 150 to 50 million years ago. Today temperatures at the bottom of the ocean are close to zero degrees.

Fossils from the past 600 million years also suggest that the earth has usually been warmer than it is today. Coal beds in Greenland and Antarctica indicate that tropical plants once grew there. The significance of such evidence is complicated by the drifting of the continents, which may have placed a given land mass in a latitude warmer than the one it is in today. Less ambiguous testimony is provided, however, by the fact that in some geologic periods, such as the Carboniferous of between 345 and 280 million years ago, coal beds were laid down all over the world.

There is no universally accepted explanation for the warmer temperatures of the past. The most likely causes, however, are continental drift and changes in atmospheric composition. Changes in the distribution of the continents could have a substantial effect on the climate because they would change the patterns of atmospheric and oceanic circulation. For example, 100 million years ago the Arctic Ocean was part of the Pacific. Winds and the ocean currents they drove might thus have kept the Arctic free of ice. As the Bering Strait closed, the freezing of the Arctic Ocean might have launched the earth into its present phase of cool climates.

Another possibility is that the atmosphere may have held far more carbon dioxide than it holds at present. Carbon dioxide is the most likely gas for this role because large quantities of it are stored in surface reservoirs from which it might conceivably have been released into the atmosphere. If the atmosphere held 100 to 1,000 times more carbon dioxide than it holds at present, the

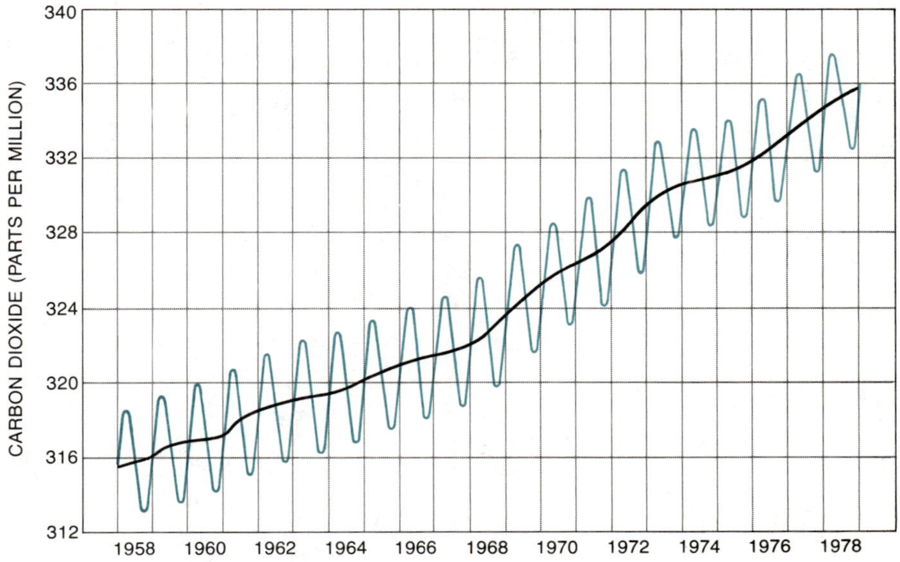

CONCENTRATION OF CARBON DIOXIDE in the atmosphere at a site near the summit of Mauna Loa on the Island of Hawaii has been measured continuously since 1958 by Charles D. Keeling of the Scripps Institution of Oceanography and the National Oceanic and Atmospheric Administration. The site was chosen because there was little local contamination of the air and because the air over this part of the globe is probably well mixed. Superimposed on the steady increase in carbon dioxide (*black*) are seasonal fluctuations (*color*) due to the uptake of carbon dioxide by Northern Hemisphere plants in the summer and the oxidation of plant tissues in the winter. The fluctuations are dominated by Northern Hemisphere plants because there is less land in the Southern Hemisphere and less seasonal change in plant life at the Equator. Measurements taken near the South Pole show a comparable increase in carbon dioxide.

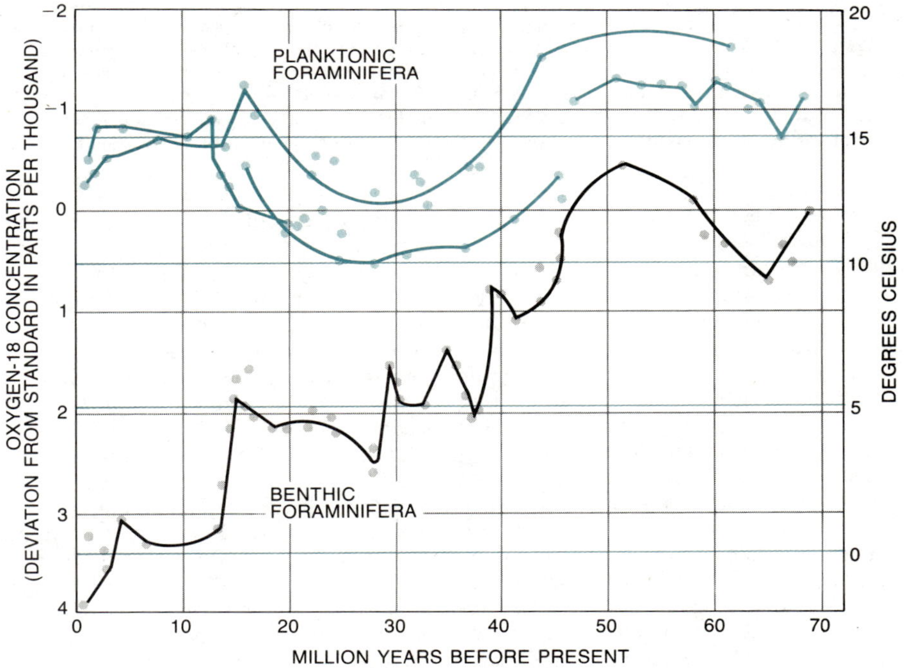

CLIMATE OF THE DISTANT PAST was characterized by higher mean global temperatures than those recorded today. One paleothermometer is the ratio of oxygen 18 to oxygen 16 in the sediments formed by the shells of the microscopic marine animals called foraminifera. Measurements of isotope ratios made by Samuel M. Savin of Case Western Reserve University show the cooling trend at high latitudes during late Cretaceous and Cenozoic times. At the end of the Cretaceous the temperature of the bottom of the ocean, which is now about zero degrees C., was about 12 degrees. Temperatures during much earlier periods were even higher.

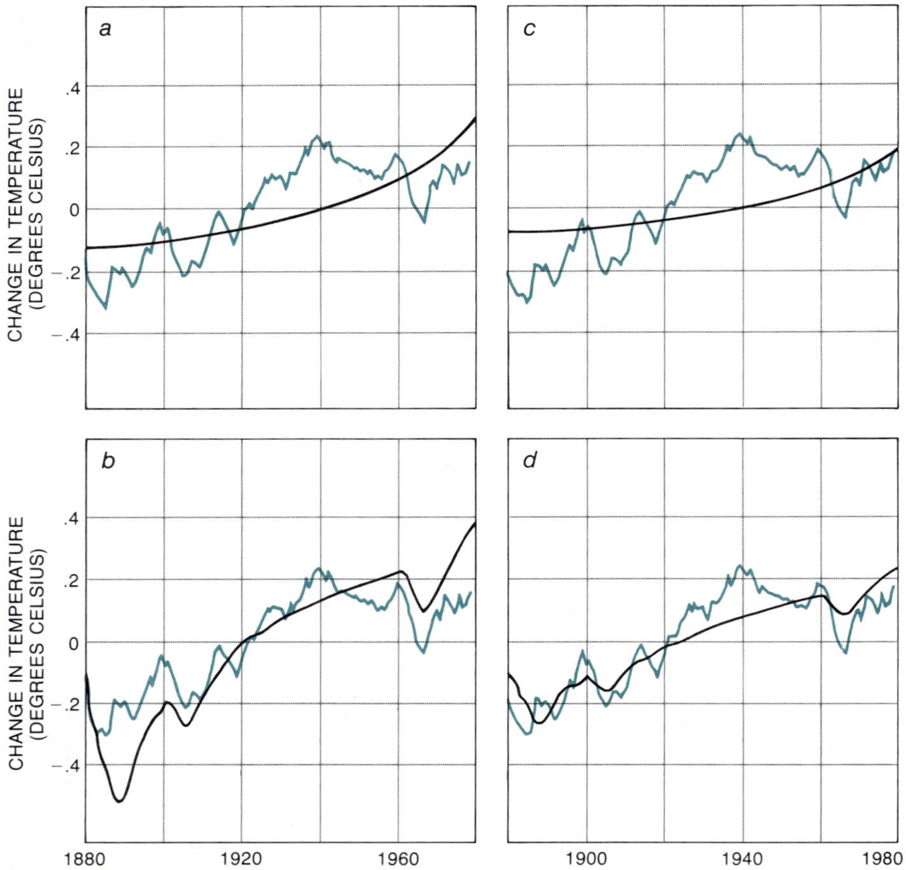

CARBON DIOXIDE AND VOLCANIC-AEROSOL LEVELS may be largely responsible for the slight warming trend in global temperature since 1880 (*color*). This period was used as a test case for a climate model (*black*) developed by James E. Hansen and his colleagues at NASA's Goddard Institute for Space Studies. Some versions of the model (*a, c*) considered only the warming due to increasing atmospheric carbon dioxide (assuming that a doubling of carbon dioxide results in a rise in temperature of 2.8 degrees C.). Other versions of the model (*b, d*) also took account of the warming due to the clearing of the stratosphere. The rise in temperature from 1910 to 1940 appears to be due largely to the clearing of the stratosphere during a period of low volcanic activity. The oceans moderate the temperature fluctuations because of their capacity to store large amounts of heat. In some versions of the model (*a, b*) only the heat capacity of the well-mixed surface layer of the oceans was considered. In others (*c, d*) the upper ocean was assumed to mix across the thermocline between it and the deeper ocean.

increased reradiation due to the carbon dioxide, together with that due to the larger amount of water vapor in a warmer atmosphere, would have maintained the surface at temperatures 20 to 30 degrees C. higher than they are today. Since the carbon dioxide in the atmosphere is currently in equilibrium with the carbon compounds dissolved in the ocean and stored in ocean sediments, however, this hypothesis implies that there were major differences between the ocean's composition 100 million years ago and its composition today. Not enough is known about the composition of either the ocean or the atmosphere of the distant past to prove or disprove this hypothesis.

The forcing mechanisms described so far are slowly varying and stimulate relatively slow responses. It is becoming clear, however, that not all changes in the climate are gradual. When a large meteoritic body strikes the earth or a large volcano erupts, significant amounts of dust and condensable gases may be injected into the atmosphere. The heavy particles fall out rapidly but the lighter ones injected into the upper atmosphere can linger for years, affecting the absorption of sunlight and the emission of infrared radiation. The consequences for the atmosphere, the ocean and the biosphere can be significant. For example, the dust injected into the atmosphere by the impact of a large meteoritic body may have caused the mass extinctions of organisms at the end of the Cretaceous period 65 million years ago. By the same token, global warming or cooling trends in the more recent past may be correlated with periods of low or high volcanic activity.

One test case for the effect of volcanic aerosols on climate is a global warming of about .4 degree C. that occurred between 1900 and 1940. It should be noted, however, that this global warming is barely significant compared with statistical fluctuations in global temperatures and the bias introduced into the measurements by urban development around many weather-observing stations. Owen B. Toon and James B. Pollack of NASA's Ames Research Center have pointed out that the warming might have been due in part to the clearing of the atmosphere after a period of high volcanic activity from 1880 to 1910. Variations in the global load of atmospheric dust with time can be estimated from a catalogue of eruptions and estimates of the size of each eruption, assuming a stratospheric residence time for particulates of about two years (derived from the dates of reports of intensely red sunsets following the explosion of Krakatau in 1883 and from the fallout rates of radioactive debris from aboveground nuclear tests).

Given an estimate of the total volume of material in the stratosphere and assumptions about average particulate size and composition based on measurements of aerosols currently in the stratosphere, the increase in the amount of sunlight reflected into space and the decrease in the amount of infrared radiation emitted into space can be estimated. The measurements indicate that the effect of the particles on incoming sunlight is greater than their effect on outgoing infrared radiation. The stratospheric aerosols therefore cool the earth. James E. Hansen and his colleagues at NASA's Goddard Institute for Space Studies have used these estimates and those of the contemporaneous increase in atmospheric carbon dioxide to test their climate model. They found that these two factors could indeed have given rise to the observed global warming between 1900 and 1940.

More recent volcanic eruptions have been studied in greater detail, so that more precise estimates of their effect on the climate can be made. It has become clear that the total volume of material ejected by a volcano is not entirely reliable as a guide to its effect on climate; the force of the eruption and the size and composition of the particulates also need to be considered. For example, the eruption of Mount St. Helens in Washington in May, 1980, was larger than that of El Chichón in Mexico in April, 1982, but lofted less stratospheric aerosol. Most of the particles ejected by Mount St. Helens were large and fell out of the atmosphere in a matter of weeks. Although less material was ejected from El Chichón, more of it reached the stratosphere and remained there.

El Chichón apparently emitted a larger quantity of sulfur than Mount St. Helens. The sulfur forms sulfur dioxide, which reacts with water vapor in the stratosphere to produce a smog of sulfuric acid droplets. Such droplets are chemically stable and settle very slowly out of the stratosphere. Predictions of the ultimate effect of the El Chichón dust cloud on the climate over the next few years are still preliminary, but the cloud is generally expected to cause a mean global cooling at the earth's surface of between .3 and one degree C.

Such atmospheric dramas are ones at which the human species is only a spectator, but it will soon be a participant. The first measurable effect of human activity on the global climate will probably be a global warming due to increases in atmospheric carbon dioxide. The amount of carbon dioxide in the atmosphere and the ocean has been increasing since the Industrial Revolution as a result of the combustion of coal and oil and the clearing of forests (both of which lead to the oxidation of carbon and the release of carbon dioxide). The preindustrial value for the amount is not known exactly, but it has been estimated at between 250 and 300 parts per million. The amount in the air over Mauna Loa on the island of Hawaii has been measured continuously since 1958 by Charles D. Keeling of the Scripps Institution of Oceanography. In the period between 1958 and 1980 atmospheric carbon dioxide showed a trend of increase from 315 parts per million to 336 parts per million.

The increase in atmospheric carbon dioxide is about half the amount of carbon dioxide estimated to have been released; the ocean acts to absorb much of the surplus. The uptake of carbon dioxide by the ocean, however, is almost impossible to measure. Carbon dioxide is not uniformly distributed in the ocean as it is in the atmosphere, so that accurate measurements at one point, such as Keeling's, would not have much meaning. Moreover, the amount of carbon dioxide in all its forms (dissolved gas, bicarbonate ion, carbonate ion and organic carbon) in the oceanic reservoir is much larger than the amount in the atmospheric reservoir, so that the changes are likely to be smaller and harder to detect. Finally, the ocean reacts with carbonate sediments such as limestone, which, at least on a long time scale, may act as an even larger sink of carbon dioxide.

Wallace S. Broecker of the Lamont-Doherty Geological Observatory of Columbia University and others have worked with models of ocean chemistry and circulation to estimate the oceanic carbon dioxide increase from the known atmospheric increase [see "The Ocean," by Wallace S. Broecker, page 69]. The rate of ocean uptake is strongly influenced by the rate of ocean mixing, both the rate of mixing in the surface layer and that of mixing between the surface layer and the deeper ocean. Broecker's model incorporated estimates of the rate of ocean mixing inferred from the dispersal of the radioactive isotopes tritium (hydrogen 3) and carbon 14 manufactured by nuclear-weapons tests. These calculations show that since the Industrial Revolution an amount of carbon dioxide approximately equal to the amount that has remained in the atmosphere has gone into the ocean. The computed total carbon dioxide increase is slightly less than, but nonetheless within the uncertainties of, the estimates of the total amount of carbon dioxide that can be accounted for by the burning of fossil fuels and the clearing of forests. Hence the estimates are at least internally consistent.

Armed with these estimates of the partitioning of carbon dioxide between the oceans and the atmosphere and estimates of future fuel combustion, one can project the increase in the atmospheric carbon dioxide into the next century and beyond. According to most projections of future energy consumption, the amount of atmospheric carbon dioxide will double by the middle of the next century and will probably double again before the peak of fossil-fuel consumption is reached.

Mathematical models can then be utilized to predict the consequences of the projected increase for the earth's climate. For example, Syukuro Manabe and Richard T. Wetherald of the Geophysical Fluid Dynamics Laboratory estimate that a doubling of atmospheric carbon dioxide (from 300 parts per million to 600 parts) would increase the mean global temperature by 2.5 degrees C. At high latitudes the temperature increase is likely to be closer to five degrees C.

Fluctuations in local temperature larger than 2.5 degrees occur daily, but such a change in the mean global temperature is significant: the change would be as great as that between the height of the last glaciation 18,000 years ago and today. The secondary consequences of the warming of the earth due to the increase in the carbon dioxide level might include a further melting of the earth's polar ice, a rise in sea level by tens of meters and more precipitation at high latitudes (perhaps offset by a widening of the subtropical arid zones, which would greatly alter the distribution of the world's water resources).

As is often the case, the greatest uncertainties in these estimates are associated with the projections of what the human species will do. This does not mean, however, that environmental scientists will sit back and await the unknown. Rather, as the environmental consequences of the human drama unfold they will keep the world informed and if necessary sound the alarm. Perhaps, on the other hand, by the next century the human species will have learned to control or influence the climate for its own benefit.

8

THE BIOSPHERE

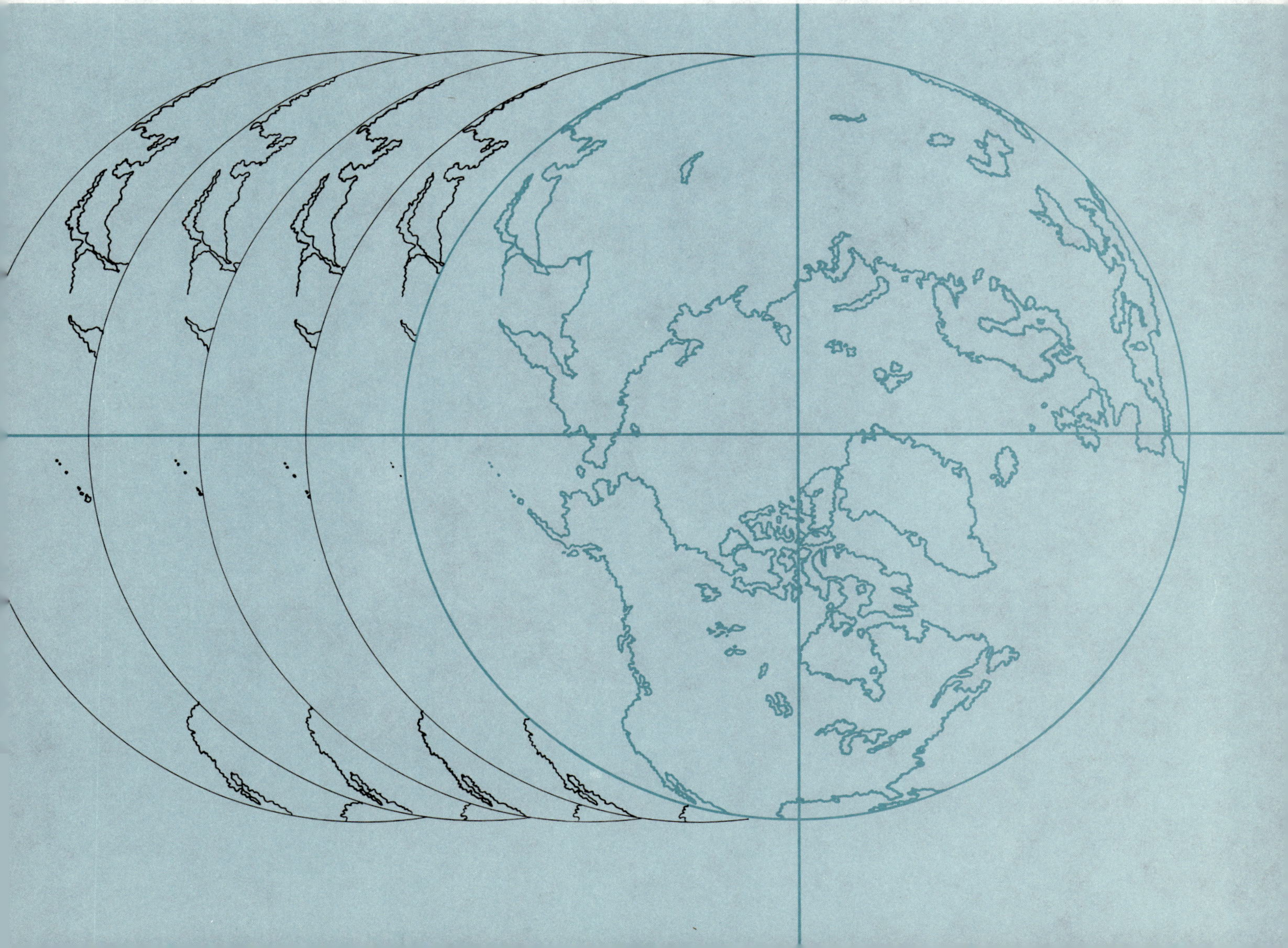

The Biosphere

by PRESTON CLOUD

The totality of microbial, animal and plant life on the earth not only is sustained by the lithosphere, the hydrosphere and the atmosphere but also has powerfully shaped their evolution

Of all the dynamic systems that make up our planet the biosphere is the last to have taken form, the most distinctive and the most interactive. Other planets and satellites in the solar system have a core, a mantle and even a crust and an atmosphere. Titan, the largest satellite of Saturn, may have seas of liquid methane, and from Mars outward through the solar system there are many icy hydrospheres. Only on the earth, however, are there structures that can replicate themselves, change into different forms by mutation and genetic recombination and transmit such changes to their descendants. Structures that can do these things are said to be living, and it is their integration everywhere on and near the earth's surface with the lithosphere, the hydrosphere and the atmosphere that constitutes the biosphere.

The diversity and interactiveness of the biosphere are rich beyond current understanding, and the potential variety within its collective gene pool is staggering. Biologists have so far catalogued some 1.5 million species of animals and .5 million species of what are generally called plants (algae, fungi and bacteria included). New species are being found and named at the rate of some 10,000 yearly. Most of the new species are insects, but the number includes such other living forms as giant deep-sea clams and numerous fossils from the still undescribed legions of plants and animals that flourished in earlier epochs. Indeed, presented with the least excuse, life teems. Responding to the variety of the earth's potential ecological niches, organisms have in geologic terms often diversified rapidly. Here we see evolution at work. Climatic and geographic isolation and such grand-scale variables as motions of the earth's lithospheric plates and the intensity of the sun's radiation provide the selective pressures that give direction to a process that otherwise seems random.

Consider the Hawaiian Islands, generated by the drifting of a lithospheric plate over a "hot spot," a rising convection current in the mantle. Kauai to the northwest, formed as the drift began, is 5.6 million years old, Oahu is 3.3 million and Maui is 1.8 million. At the southeast end of the chain the big island, Hawaii, which is still growing, emerged only 700,000 years ago. In that brief 5.6 million years of earth history the descendants of some random fruit-fly migrants to the new islands have so diversified that 25 percent of the world's total of some 2,000 species of *Drosophila* are found there and there alone. Over the same span in the island chain more than 1,000 separate species of land snails evolved in the isolated valleys that descend the volcanic slopes. And an entire new family of birds—the honeycreepers, numbering some two dozen species—has arisen from unknown finchlike ancestors, borne to the islands on some chance wind.

In the long history of the biosphere, however, the most important event by far was the origin and early diversification of life: the beginnings of biospheric interaction with the earth's surface. Indeed, I shall not be much concerned here with the fascinating diversity and complexity of the planet's living structures or with their detailed evolution. The focus instead will be on the far-reaching ways in which life has modified the history and the surface features of our world and has in turn been modified over the four billion years that have elapsed since the earth's surface first became habitable.

Today few places on the earth's surface, or below it for some tens of meters, or above it to a height of nearly 10 kilometers, are now so hot or so cold or so dry or so polluted or so exposed to ionizing radiation that they are truly devoid of life. This was not always the case. Nor was the major work of the biosphere done by its currently most conspicuous constituents: the familiar multicellular plants and animals of field, stream and sea. In the course of the biosphere's history its geochemically most influential members have always been those of the microbial realm, the humble and morphologically simple but biochemically diverse and adaptable masses of the minute.

The interaction of the biosphere, the hydrosphere and the atmosphere with one another and with the earth's outer crust is not only universal but also continuous, a matter of cycles within cycles within cycles. Animals such as worms, ants, moles and gophers till the soil until their own remains are recycled by decay bacteria and other scavengers. Meanwhile other biospheric agents, including such odd animal-plant symbionts as termites and their hind-gut bacterial helpers, recycle much of the plant kingdom, simultaneously adding large inputs to the methane and carbon dioxide components of the atmosphere. Every year living plants transfer to the atmosphere some 400,000 tons of organic volatiles, some of them incorporating metals.

A good third of all chemical elements are recycled biologically, and some living organisms are capable of concentrating comparatively rare elements far above their background abundance. All organisms utilize and recycle not only hydrogen, oxygen, carbon and nitrogen but also phosphorus, sulfur, calcium, potassium, magnesium, sodium, iron, manganese, cobalt, copper and zinc.

GREAT BARRIER REEF, in the vicinity of Cape Melville (*upper left*) in northeastern Australia, is seen in the Landsat image on the opposite page. The reef is a vivid example of how processes of the biosphere interact on a large scale with those of the lithosphere, the hydrosphere and the atmosphere. The activities of diverse life forms, notably corals and algae, have built this 2,000-kilometer natural breakwater extending from the Tropic of Capricorn northward to Torres Strait off New Guinea. The reef parallels the Queensland coast at distances ranging from 15 to 150 kilometers, and the waters between the outer reef and the shore hold many other reef structures. North is toward the top. The area shown is 60 kilometers wide.

OLDEST SEDIMENTARY ROCK, from the Isua area in southwestern Greenland, was deposited some 3.8 billion years ago. The specimen is from what is called a banded iron formation (although it lacks the reddish color found in the alternating dark stripes of other such rocks). The stripes are thinly stratified layers, now folded and stretched. Formations of this kind were among the oxygen "sinks" that kept free oxygen nonexistent or scarce in the early atmosphere of the earth. The rhythmic banding of the specimen, together with a ratio of carbon isotopes reminiscent of life processes, suggests that microbial organisms may already have been at work.

RED SHALE, estimated to be about two billion years old, overlies Cambrian granites near Lac Cambrien in Quebec that are roughly 2.5 billion years old. Among the oldest "red beds" known, it underlies the youngest banded iron formations. Red beds mark a transition from essentially anoxic conditions to the inauguration of an atmosphere where some oxygen was always present.

Most organisms utilize chlorine in some way, and about a dozen other elements are exploited for particular biological functions. Biological processes are responsible for massive concentrations in the earth's crust of silicon, iron, manganese, sulfur and carbon. Microbes can thrive on or in the presence of such corrosive substances as sulfuric acid, carbolic acid and hydrogen sulfide.

Rock weathering, soil formation and many sedimentary rocks are important partial products of microbial and other biological processes. The oxygen in the atmosphere is overwhelmingly from photosynthesis by green plants, which also segregate carbon in soils and sediments. The domestic animals that provide amino acids essential to human nutrition would not be able to function without their digestive bacteria, nor would we. And many of the plants on which animals feed depend on the nitrogen-fixing activities of bacteria and their first cousins the blue-green algae or proalgae—carryovers from the earth's earliest history. The biosphere profoundly and pervasively affects the rest of the earth's surface and is in turn acutely responsive to feedbacks from other spheres of activity.

The appearance of life was thus a dramatic innovation on the barren prebiotic planet, introducing an unprecedented and far-reaching geologic process. Although one cannot say with confidence just how and when life first arose on the earth, conjecture is now limited by evidence on likely abiotic sources for the initial organic but nonliving macromolecules and the autocatalytic processes that might have organized them into components of the earliest cells. What can be said with some confidence is that the planet's initial life forms were biochemically simple, unicellular (or noncellular), probably spheroidal in shape and dependent on extracellular sources of nutrients. If such objects were eventually to be found as fossils, they probably could not be distinguished on morphological grounds from similar objects of nonbiological origin.

It is also possible to estimate within limits when and under what conditions the protobiosphere first began and even to reconstruct some elements of its subsequent history and interactions with the nonbiological world. The evidence on biospheric processes in earth history belongs to the broad field of biogeology. Its data include the remains of the microorganisms found, sedimentary structures produced by their activities, the geochemistry and isotopic composition of biologically significant elements and the biogeochemistry of organic substances and products. The oldest two major signposts are both in Western Australia. One of them, near a place so isolated that it is called North Pole, is

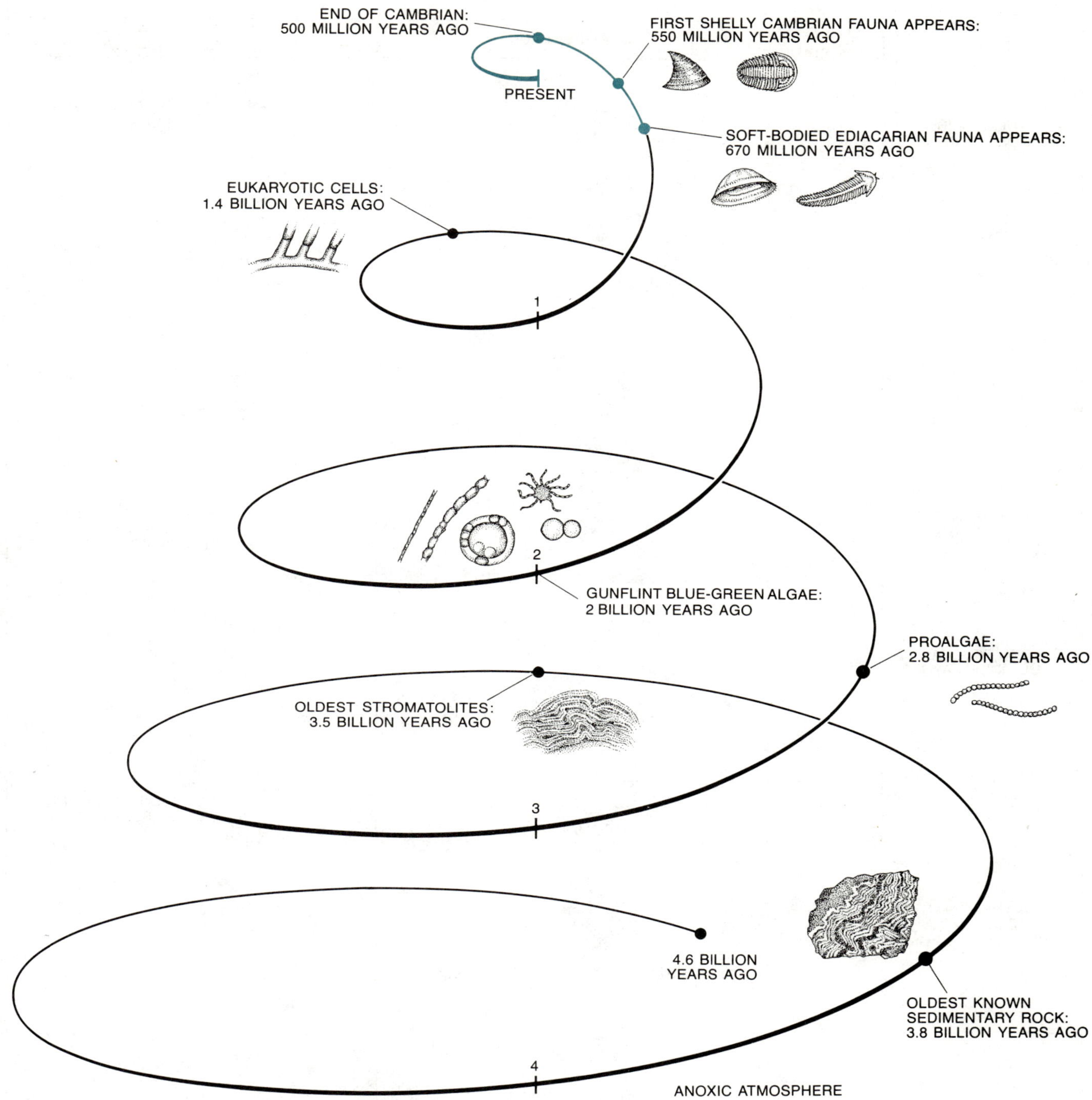

ASCENDING SPIRAL schematically represents the earliest phases of the earth's biogeologic history, beginning soon after the solar system formed some 4.6 billion years ago, when the planet was inhospitable to life and a direct geologic record is nonexistent. By 3.8 billion years ago the developing atmosphere probably consisted mainly of carbon dioxide, water vapor, nitrogen, carbon monoxide, hydrogen sulfide and hydrogen. Free oxygen was absent or trivial and transient. The temperature of the earth's solid surface had by then fallen from about the melting point of iron to a mean temperature between the boiling and freezing points of water. Even though the young sun shone with only 60 to 70 percent of its present intensity, a life-supporting range of temperatures was made possible by the "greenhouse effect" of an atmosphere rich in carbon dioxide. Some 300 million years later sedimentary structures called stromatolites appear; they resemble deposits formed today by the photosynthetic and sediment-binding activities of blue-green algae. After about another 700 million years sedimentary rocks being deposited in what is now Western Australia included tiny filamentous structures that are convincingly microbial and may be proalgae. By two billion years ago, a date on the third loop of the spiral, accumulating silica gels of a Lake Superior iron formation were trapping filaments, cells and more complex structures that are unquestionably microbial. Some show two or three cell types similar to the living blue-green alga *Nostoc,* a typical prokaryote. (Prokaryotic cells lack the nucleus characteristic of eukaryotic cells, the building blocks of all higher forms of life.) By then the oxygen content of the hydrosphere and the atmosphere is estimated to have reached about 1 percent of its present level, allowing a tenuous ozone screen to develop and shield the earth's surface from ultraviolet wavelengths inimical to living organisms. The oldest persuasively eukaryotic cells entered the fossil record about 1.4 billion years ago. Their appearance was announced by a widespread increase in the average diameter of the cells. By about 670 million years ago a remarkable and diverse group of soft-bodied aquatic animals, the first known metazoans, entered the fossil record in the sedimentary deposits of the Ediacara Hills in South Australia. Their lack of a shelly covering suggests that the oxygen level was by then near 7 percent of the present level. Their appearance marks the end of earlier divisions of geologic time and the start of the present eon, the Phanerozoic. This is the period of manifest rather than cryptic animal life that appears in color on the fifth and final loop of the spiral. Soft-bodied and shelly metazoans are preserved as fossils in Cambrian sediments, deposited between 550 and 500 million years ago.

EVENT	BILLIONS OF YEARS AGO	MANIFESTATION	OXYGEN (PERCENT)	COLLATERAL EVENTS AND CONSEQUENCES
8. FULLY OXIC CONDITIONS	.4	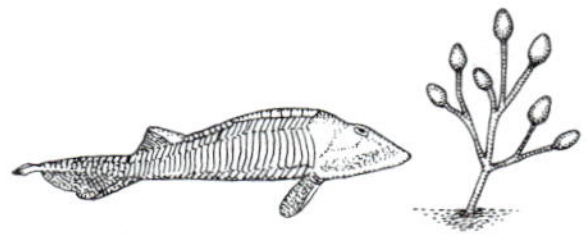 LARGE FISHES, FIRST LAND PLANTS	100	BIOSPHERIC EVOLUTION TOWARD PRESENT WORLD STATE
7. SHELLY METAZOANS APPEAR	.55	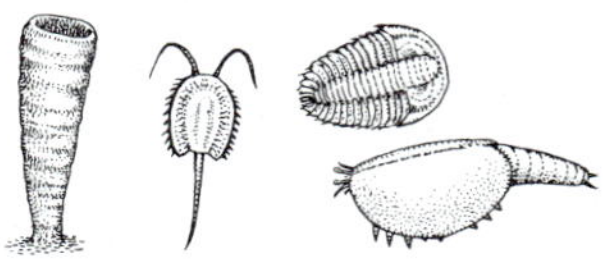 CAMBRIAN FAUNA	~10	BURROWING HABIT BEGINS; SUBSEQUENT EVOLUTION
6. METAZOANS APPEAR	.67	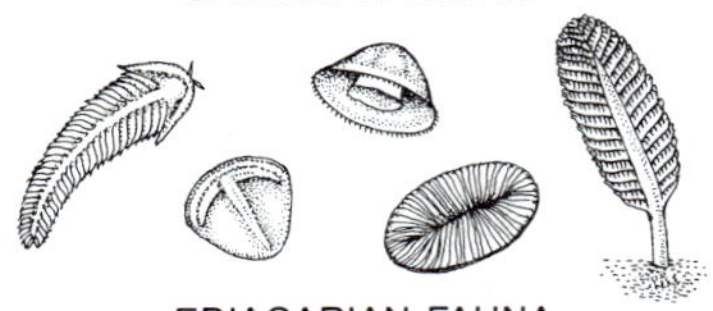 EDIACARIAN FAUNA	~7	PHANEROZOIC EON BEGINS; METAZOAN FOSSILS AND TRACKS
5. FIRST EUKARYOTIC CELLS FOUND	1.4	 CELLS LARGER IN DIAMETER	>1	RED BEDS PROLIFERATE, MUTICELLULAR ORGANISMS; MITOSIS, MEIOSIS, GENETIC RECOMBINATION
4. OXYGEN-TOLERATING BLUE-GREEN ALGAE	~2.0	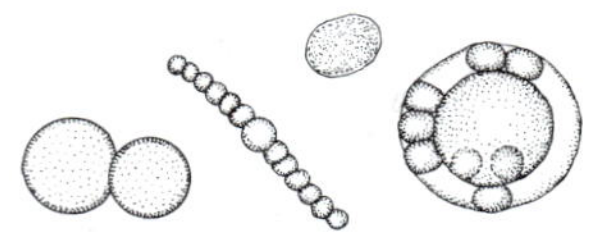 ENLARGED, THICK-WALLED CELLS AT INTERVALS ON ALGAL FILAMENTS	~1	OXIDATIVE METABOLISM, OZONE SCREEN; OLDEST RED BEDS OVERLAP WITH YOUNGEST BANDED IRON FORMATIONS
3. PHOTOAUTOTROPHY, PROBABLY RELEASING MOLECULAR OXYGEN	>2.8	 STROMATOLITES, PRECURSORS OF BLUE-GREEN ALGAE	<1	CHLOROPHYLL *a* AND CYTOCHROME *b* SYNTHESIZED; BANDED IRON FORMATIONS AND OTHER OXYGEN SINKS; GREEN-HOUSE EFFECT DIMINISHED
2. AUTOTROPHY (METHANOGENESIS?) (SULFUR OXIDATION?)	>3.5	 STROMATOLITES, SULFATE, LIGHT CARBON	(ANOXIC)	BIOSPHERIC CONTINUITY
1. ORIGIN OF LIFE	(~3.8?)	 LIGHT CARBON	(ANOXIC)	EVOLUTION OF BIOSPHERE BEGINS

EIGHT MAJOR STEPS in the early evolution of the biosphere are listed in this table, beginning at the bottom with the origin of life on the earth about 3.8 billion years ago (*1*). This date is suggested by the diminution of the isotope carbon 13 with respect to the lighter isotope carbon 12 ("light carbon" in the table), a ratio characteristic of life processes. A few hundred million years later (*2*), although oxygen was not yet permanently present in either the atmosphere or the hydrosphere, the appearance of stromatolites implies that some microbial form of self-feeding organism had evolved. Perhaps 800 million years after that (*3*) additional stromatolites and the fossilized remains of what appear to be blue-green algae or their precursors lead to the conclusion that oxygen-releasing photosynthesis might have begun, presenting the problem of how life was to be shielded from this corrosive element. The appearance of banded iron formations, whose sediments served as large oxygen sinks, may have postponed the problem. About two billion years ago (*4*) blue-green algae whose chains of tiny cells included thick-walled ones such as those shielding nitrogenase enzymes from free oxygen today indicate that oxygen was accumulating in the hydrosphere and suggest the appearance of oxidative metabolism as a superior energetic pathway. The first continental red beds appear at about this time, implying development of an atmosphere that contained oxygen permanently. Between then and 1.4 billion years ago (*5*) eukaryotes, generally larger cells whose DNA is shielded within a membrane-bounded nucleus, signal the evolution of three characteristic traits: mitosis (in which rodlike chromosomes line up and divide in a spindle), meiosis (in which the chromosomes are divided in half for reproductive purposes) and the advanced evolutionary process of genetic recombination. Between 670 and 550 million years ago fossils found at sites on five continents (*6*) present the outcome of the continuing proliferation of eukaryotes: the Ediacarian fauna, multicellular aquatic animals that mark the start of the Phanerozoic eon, in which all life on the earth still lives. By 550 million years ago (*7*) the Cambrian fauna, a worldwide succession of ancient organisms, including the first (or almost the first) shelled invertebrates, enters the fossil record. It is estimated that by then the supply of oxygen had reached about 10 percent of the current level. Some 150 million years later (*8*) the probability that the present level of oxygen was being approached is implied by the presence of large and active fishes in the sea and plants and invertebrate animals on the land. Their descendants went on to fill all the planet's ecological niches.

found in rocks about 3.5 billion years old (the Warrawoona Group) in the form of inferentially biogenic sedimentary structures called stromatolites (and microfilaments that may not, however, be contemporaneous with the enclosing sediments). The second signpost, from 2.8-billion-year-old strata of the Fortescue Group, also in Western Australia, contains the oldest fossil paleomicrobiota that I find to be convincing both as to their biogenic nature and as to their contemporaneity with the rocks in which they are found.

A third signpost is firmly planted in two-billion-year-old siliceous rocks called chert on the north shore of Lake Superior. They are a part of the Gunflint Iron Formation, which contains a variety and abundance of demonstrably biogenic and unequivocally contemporaneous microbial fossils that compare favorably with the richest modern microfloras. Indeed, the Gunflint microbiota contains the oldest fossils known to display a clear differentiation into two or more types of cell. A common filamentous microfossil closely resembles living freshwater blue-green algae of the genus *Nostoc*. Others resemble budding bacteria. Their presence in this formation, together with their resemblance to living organisms, implies a continuity of biological functions from two billion years ago to today.

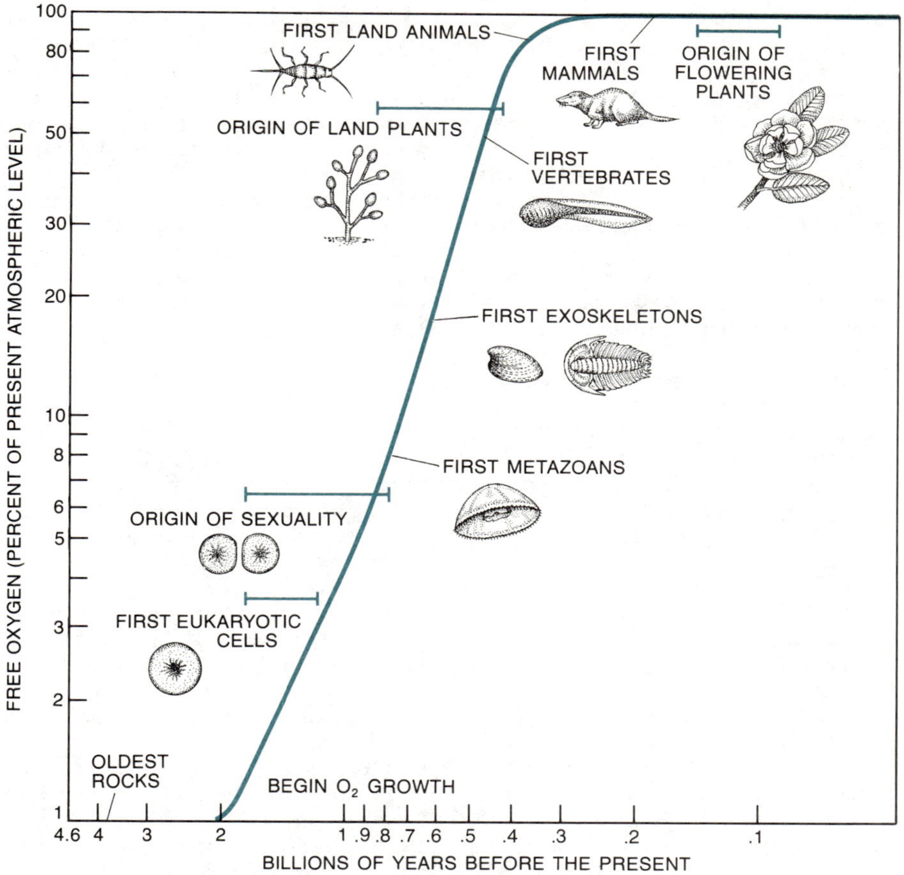

ENRICHMENT IN OXYGEN of the earth's originally anoxic atmosphere is traced on this double logarithmic graph. The abscissa measures billions of years before the present; the ordinate shows the estimated increase of molecular oxygen from about 1 percent of its present level to what it is today. Leaders indicate the timing of key events in the evolution of the biosphere as they relate to the evolution of the atmosphere. In the case of events where there is uncertainty about the time of origin a range bar (*color*) indicates the possible limits. For example, microbial land plants of the simplest kind might have appeared at or before the beginning of the Phanerozoic eon, but spores that have been interpreted as being those of true vascular plants did not appear until the late Ordovician, some 440 to 430 million years ago. In turn, the oldest organisms generally accepted as being land plants in the modern sense are of late Silurian age, perhaps 420 to 415 million years old. Scorpionlike arachnids of about the same age are known. Primitive insects do not appear until the Middle Devonian, some 380 million years ago.

Earlier records of microbial life are less conclusive. The record of the 2.8-billion-year-old Fortescue Group consists of filamentous chains of cell-like entities that show no cellular differentiation but otherwise resemble certain living blue-green algae. The 3.4-to-3.5-billion-year-old Warrawoona rocks at the locality from which the best material was collected show three separate sets of postdepositional fracturing, have undergone changes in their initial chemical state and are heavily infiltrated by secondary iron. The best-preserved and most abundant microbial forms in these rocks resemble the flat, twisted stalks of the modern iron bacterium *Gallionella ferruginea*. Evidence that they are the same age as the sediments in which they are found is not convincing. In rocks that are considered to be the same age as those containing this microflora, however, there are laminated and undulating pseudocolumnar and domal stromatolite structures like those being built in shallow waters today by colonies of blue-green algae. These very old stromatolites are presumptive evidence (not proof) of a microbial presence in Warrawoona time, perhaps the progenitors of later blue-green algae or their bacterial cousins.

Even more indirect evidence for a still earlier signpost is found in carbonaceous sediments of the Isua area in southwestern Greenland. The carbon-isotope ratios of these 3.8-billion-year-old rocks show a depletion of the isotope carbon 13 with respect to the lighter isotope carbon 12, a common manifestation of biospheric activity. It would be consistent with such evidence to postulate the presence of life even that early. The fractionation observed, however, is not conclusive as to its source. It could be the work of aerobic or anaerobic photosynthesizing organisms, of methane-assimilating bacteria (methanogens) or conceivably of some nonbiological process.

What was the terrestrial environment like in those early times? So far no rocks have been found on the earth older than those at Isua, although much older rocks are known on the moon and in the form of stony meteorites. Considering the evidence for a high rate of meteoritic and cometary bombardment throughout the early history of the solar system, however, it can safely be inferred that up to about four billion years ago the earth was inhospitable to life as we know it. Initial surface temperatures near or above the melting point of iron, the absence of both an atmosphere and a hydrosphere and intense, unshielded solar radiation would all have contributed to the harshness of the primordial environment. Like life, however, the earth evolves, and these harsh conditions changed.

A free interpretation of the geologic record in the light of basic biological principles suggests the following surface milieu some 3.8 billion years ago. The developing atmosphere at that time was anoxic: it lacked any permanent free oxygen. Its main gases were most likely carbon dioxide, nitrogen, water vapor, carbon monoxide and perhaps hydrogen sulfide. Traces of hydrogen, hydrochloric acid, ammonia and methane would probably also have been present. There were no primary sources of oxygen, neither ordinary molecular oxygen (O_2), singlet or atomic oxygen (O) nor ozone (O_3). Whatever transient free oxygen may have formed as a result of the photolytic breakdown of water vapor and carbon dioxide was rapidly consumed by chemical recombination, by reduced gases expelled from the earth's interior and by a variety of other oxygen sinks. The developing hydrosphere was probably salty, although probably less salty than modern seas as a result of dilution by water of cometary origin. The ancient sedimentary rocks of Isua ensure that the mean surface temperature of the planet as far back as we have any record was above the freezing point of water and below the boiling point. The freezing suggested by the astronomers' "faint early sun" was presumably counteracted by the "greenhouse" effects of atmospheric carbon dioxide, water vapor and ammonia.

How could life have arisen in such a setting? That calls for a source or sources of energy, a templating structure or mechanism to explain the chirality, or asymmetric "handedness," of amino acids and sugars, a local concentration of organic macromolecules and catalytic effects to hasten and direct the process. Experiment, observation and reflection support the view that energy for prebiotic and early biochemical reactions was available from solar ultraviolet radiation and possibly other sources (for example lightning or chemical reactions). The templating and concentration of abiogenic organic molecules may have been favored by the properties of asymmetric minerals such as clays, or perhaps by polarized light. And autocatalytic effects were surely possible in the form of the frequent day-night freeze-thaw cycles of the rapidly rotating early earth, of iron and magnesium compounds, of dehydration condensations (the linking of molecules by the loss of combined water) and of chemical selection. Whatever the details may have been, the origin of life was an epic event. Given the record of biospheric interactions, nothing at or near the earth's surface could ever have been the same again.

From such an unpromising, oxygen-depleted start how might the initial biosphere have become the biosphere of today? The central and most general driving force was natural selection: a response to changing ecological challenges and opportunities. Among these I shall stress the role of oxygen.

Oxygen, primarily in its molecular form, was to the early biosphere what, by analogy, nuclear power is to the biosphere of today: pregnant with potentialities and cursed with contradictions. The biosphere is a huge metabolic device for the capture, storage and transfer of energy. It carries out these metabolic functions in two modes: fermentation and respiration, both involving the enzymatic conversion of glucose into energy through intermediate steps. One group of organisms, all bacterial, operates by fermentation only. The metabolism of most other organisms, mainly the ones we call higher but also some microbes, is respirational. To the initial fermentative step called glycolysis, respirational metabolism adds the set of reactions known as the citric acid cycle, which through the process of oxidative phosphorylation increases the amount of energy generated 16-fold. In the citric acid cycle glucose is converted into biological energy in the form of adenosine triphosphate (ATP).

From this it is inferred that fermentation is the primitive form of metabolism and respiration the derived form. Only thus can one account for their near-correspondence, with the separation of life into two primary categories, successional in time and built from dramatically different cell types: (1) the prokaryotes, which lack a cell nucleus and are mainly fermentative, and (2) the eukaryotes, which have a nucleus and are mostly obligatory aerobes. So also can we best account for the superimposing of the citric acid cycle on fermentational beginnings. The blue-green algae are functionally intermediate prokaryotes, some of which can switch off oxidative phosphorylation and function on fermentation alone. The first green-plant photosynthesizers, they have not completely cut their ties with their anaerobic bacterial ancestors.

The principal driving forces behind the interaction of the biosphere with the lithosphere, the hydrosphere and the atmosphere are therefore the production of glucose, utilizing external sources of energy, and the metabolic conversion of glucose into ATP, particularly as enhanced by oxidative phosphorylation in the citric acid cycle. When the early

GUADALUPE MOUNTAINS of Texas, near the New Mexico border, are capped by a great bluff, here covered by clouds, that was built mainly by sponges and algae. It is part of a reef, 640 kilometers long, that gradually arose in late Permian times, 240 to 230 million years ago. Most of this Paleozoic reef is now buried under younger sediments. The cloud-capped peak, El Capitan, reaches 8078 feet.

blue-green algae or proalgae first left unmistakable traces in the form of the Gunflint microbiota almost two billion years after life may have originated on the earth, the atmosphere probably had only about 1 percent of its present level of molecular oxygen. This meant, however, that the initial oxygen sinks were becoming neutralized and that oxygen was now accumulating in the hydrosphere and beginning to escape into the atmosphere. Oxidative phosphorylation was surely by then a fact of life.

It is now appropriate to consider more fully how the evolution of the biosphere may have been linked with the other great spheres of activity that make up the dynamic earth. Although the evidence is still dauntingly incomplete, it is sufficient to formulate a consistent preliminary account of probable early linkages among these spheres. In addition to the increase of oxygen in the hydrosphere and the atmosphere I shall briefly consider the evolution of the eukaryotic cell, foreshadowing the development some 700 million years later of the Metazoa: multicellular animal life. Thereafter plate tectonics had a profound influence on the evolution of the Metazoa and land plants. Here, however, I shall emphasize microbial processes, later assisted by the higher algae and land plants.

What really needs explaining about atmospheric oxygen is not its scarcity in the beginning but its abundance in the end; from anoxic in the beginning, to perhaps 7 percent of the present atmospheric level when metazoans first appear, to near the present level by mid-Paleozoic times. This much, at least, is clear: omitting minor amounts of photolytic oxygen, the increase in hydrospheric and atmospheric oxygen following the filling of the major oxygen sinks depended on the sedimentary segregation of an equivalent amount of carbon. To show why, I shall simplify the photosynthetic equation to read $CO_2 + H_2O \rightleftharpoons (CH_2O)_n + O_2$. In order to create and maintain the earth's oxygen-rich atmosphere the C in the CH_2O (or glucose made from it) must be buried in the sedimentary column faster than the O_2 is consumed by recombination or by the oxidation of previously buried carbon and new reduced volcanic gases, a cycle that today takes only three million years. In other words, atmospheric oxygen is preponderantly the result of a lag in the geochemical cycling of the products of photosynthesis.

Since human beings, like most other eukaryotes, depend for their biological energy on respiratory metabolism involving the oxidation of pyruvate derived from glucose, they are likely to think of molecular oxygen as being essential to life. On the contrary, oxygen is poisonous to all forms of life in the ab-

BIOSPHERIC HISTORY over a span of nearly two billion years is preserved in these strata, which rise 1,340 meters from the desert floor. At the bottom are sandstone red beds that were formed in Upper Proterozoic times, 1.1 billion years ago. Just below the peak are other red beds, formed in Upper Paleozoic times, from 300 to 270 million years ago. At the crest of the peak is a formation of marine sediments that accumulated when the area was covered by sea at the end of the Paleozoic. The peak, Comanche Point, is a landmark of the Grand Canyon.

sence of the enzymes needed to reduce destructive by-products of oxidative metabolism such as hydrogen peroxide and superoxide. Nature has also had to be clever in shielding from oxygen nuclei and other critical sites in the living cell and in designing oxidative pathways that work by the removal of hydrogen rather than by the addition of oxygen.

In fact, what oxygen is essential to is not the life process itself but a high level of metabolic energy. The need of eukaryotic organisms for oxygen is entirely for the production of the basic energy-transferring molecule ATP. If nonoxidative processes yielded as much ATP as oxidative processes did, there would be no demand for metabolic oxygen. It is equally certain that prebiotic chemical evolution leading to an initial stock of organic macromolecules could not have taken place in the presence of free oxygen in any form. The reactions would not go and their products would not survive. Nor could early life have survived in the presence of free oxygen before enzymic defenses against it evolved, except under vanishingly low and transient oxygen concentrations.

These indirect but well-founded biochemical conclusions are reinforced by geochemical and paleomicrobiological evidence for an initially oxygen-free earth. The record shows that the few believable fossils known before about two billion years ago were spheroidal forms and filamentous chains of cells so small and of such simplicity that they imply a prokaryotic nature and therefore a limited tolerance to oxygen. With the presence of such ample oxygen sinks as sulfide gases, sulfide minerals, ferrous iron and reduced gases from the extensive Archean volcanic belts of 2.5 billion years ago and more, these tiny organisms were under no selective pressure to acquire defenses against oxygen. Indeed, no such pressure was to arise until after the evolution of photosynthesis by blue-green algae, with its release of molecular oxygen, and after the final saturation of the major oxygen sinks.

The oldest paleontological evidence that biological oxygen was beginning to accumulate in a previously anoxic hydrosphere and thus also to escape into the atmosphere comes from the Gunflint Iron Formation of roughly two billion years ago. The filamentous microbial species *Gunflintia minuta* shows occasional enlarged cells that strikingly resemble the thick-walled cells found at intervals along the filaments of such living blue-green algae as *Nostoc*. Among such species the thick-walled cells, called heterocysts, shield the enzymes required for nitrogen fixation from the oxygen that would otherwise destroy them. Such cells lack photosynthetic pigments and neither produce nor tolerate oxygen. The similarity is too close to be fortuitous. The Gunflint heterocysts imply that two billion years ago oxygen in the atmosphere had reached levels in excess of the level that now inactivates the nitrogenase enzymes. They further imply that at that time, if not earlier, there existed something very much like living nostocacean algae. Such organisms were presumably capable of splitting the water molecule as a source of the oxidative energy needed to add the citric acid cycle to an entirely fermentative metabolism.

How might the first blue-green algae, or their predecessors, have acquired such a characteristic? How might they have limited the free oxygen generated as a by-product of this activity to levels that could be managed by primitive oxygen-mediating enzyme systems? Answers to these questions are suggested by representatives of the half-dozen genera of modern blue-green algae that can switch from aerobic metabolism to anaerobic. These species prosper only in the presence of hydrogen sulfide, a reducing gas that keeps the pressure of ambient oxygen low. They have a weak tolerance for oxygen and are able to switch from water to hydrogen sulfide as a source of energetic electrons for biochemical reactions. The product of such anaerobic photosynthesis is two sulfur atoms rather than one molecule of oxygen; the sulfur atoms then become available for conversion into sulfate ion by sulfur bacteria.

Some mutant photosynthetic sulfur bacterium may have been the first to acquire reaction-energizing electrons by splitting water molecules instead of hydrogen sulfide, while at the same time retaining its ability to use hydrogen sulfide as an alternative energy source. The success of such a mutant would have been assured when its ability both to generate and to tolerate the by-product, oxygen, left it adaptively superior to microbial competitors that were exclusively anaerobic. One can imagine further mutations leading to enzymatic protection against higher oxygen concentrations while retaining (but eventually losing) access to hydrogen sulfide as an emergency energy source.

Organisms of this kind could have been responsible for unusual rock formations that were widely laid down in Archean and early Proterozoic times up to about two billion years ago. These were the finely laminated siliceous banded iron formations. Under nearly oxygen-free conditions soluble ferrous iron could have been episodically dispersed over large areas, functioning as an oxygen buffer and stimulating the growth of oxygen-producing photosynthetic microbes of limited oxygen tolerance, as hydrogen sulfide does today. Such proto-blue-green algae, in turn, could have supplied the oxygen necessary for the episodic precipitation of ferric and ferro-ferric oxides (hematite and magnetite) to make the microlaminated siliceous banded iron formations. The episodicity observed may reflect seasonal microbial blooms or episodic upwelling of ferrous iron from anaerobic basins or both. (This is not, however, to argue that *all* iron formations are so produced.) As this kind of chemical balance drew to an end the level of oxygen in the hydrosphere would have increased and so would the leakage of oxygen into the atmosphere.

The progression from a mainly anoxic hydrosphere and atmosphere to weakly oxic ones about two billion years ago is supported by two other lines of geochemical evidence. The first is the widespread presence in Africa and the Americas of the easily oxidized mineral uraninite in river sands older than about 2.3 billion years. Such extensive accumulations of so readily oxidized a mineral in stream deposits would have been improbable beneath a substantially oxic atmosphere. The other line of evidence is the near-limitation of the banded iron formations to rocks older than about two billion years. Among rocks younger than two billion years are the oldest conspicuous "red beds": sands mostly of continental origin colored by ferric oxide. Thus the banded iron formations imply a generally anoxic hydrosphere with episodes of oxidation; the red beds imply an oxidative atmosphere (and hydrosphere).

What might the level of atmospheric oxygen have been two billion years ago? Several considerations suggest that it was about 1 percent of the present level. Above that level enough ozone can form to block the harsher ultraviolet radiation of the sun. Below that level optional anaerobes can switch from aerobic metabolism to anaerobic. The apparent presence of eukaryotes by about 1.4 billion years ago implies that a level of oxygen 1 percent of the present level had been attained earlier. Uraninite in riverine sandstones up to 2.3 billion years ago indicates that at the time such a level had not yet been reached. Accordingly a continuing atmospheric level of oxygen above 1 percent is approximately bracketed between those numbers. The general transition from banded iron formations to red beds about two billion years ago suggests the passing of some threshold level of oxygen about then. Was that the first attainment on a sustained basis of 1 percent of the present atmospheric level of oxygen? Quite likely.

If the most important event in the evolution of the biosphere was the first appearance of life, it is run a close second and third by the appearance of chlorophyll *a* as the mediator of oxygen-producing photosynthesis sometime before two billion years ago and the appearance of the eukaryotic cell, with its characteristic mitotic cell division, between two and 1.4 billion years ago. Whereas in the prokaryotic cell the DNA forms a single long chromosome that is folded irregularly throughout the cell, in the eukaryotic cell the chromosomes are numerous, rodlike and shielded within a well-defined, membrane-bounded nucleus. In the process of mitosis the chromosomes cluster into a central spindle and split into pairs before each cell division. Mitosis depends on the contractile properties of the protein actomyosin, which cannot form in the absence of oxygen. The more advanced steps in the synthesis of sterols, fatty acids and the fibrous protein collagen, leading to muscles and metazoans, are also dependent on a sufficient level of oxygen.

Still, it remains unsettled exactly how either eukaryotic cells or the metazoans arose. Part of the story of the origin of the eukaryotic cell surely involves endosymbiosis: the engulfing of one organism by another, giving rise to cell organelles such as the mitochondrion and the

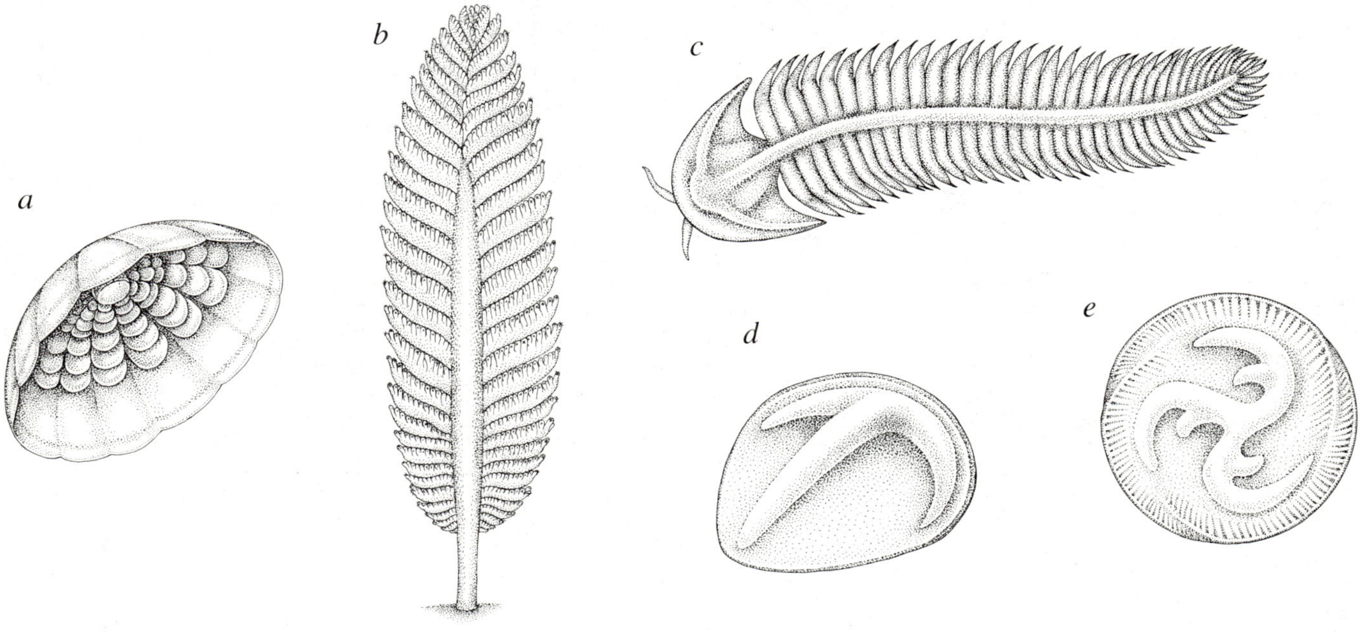

PRECAMBRIAN ANIMALS, once best known from the Ediacara Hills of South Australia, have now been found on five continents. The most abundant animals resemble jellyfish (*a*) or other modern coelenterates such as the coral-like ones known as sea pens (*b*). Others, such as *Spriggina* (*c*) resemble naked arthropods and annelid worms. Still others, such as *Parvancorina* (*d*) and *Tribrachidium* (*e*), resemble no other known animal. All probably took their supply of oxygen from the surrounding water by absorption through their epithelia.

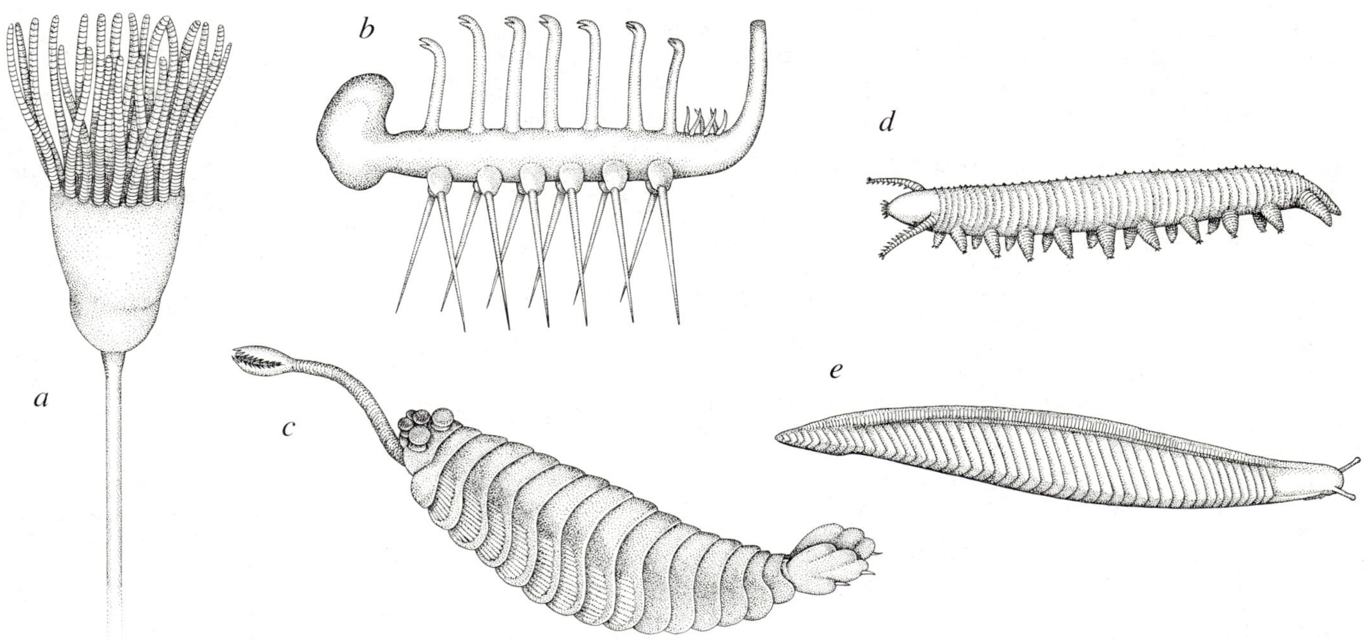

CAMBRIAN ANIMALS, far more abundant than the preceding Ediacarian ones and commonly protected by an exoskeleton, also included soft-bodied species that are best preserved in a Middle Cambrian formation at Mount Robson in Canada. By this time representatives of three new and now extinct phyla had appeared: the immobile *Dinomischus* (*a*), a stilt-legged scavenger, *Hallucigenia* (*b*), and the predaceous *Opabinia* (*c*). A fourth newcomer, *Aysheaia* (*d*), resembles the living onychophoran *Peripatus*. Probably the most advanced of the group, *Pikaia* (*e*), is interpreted as the sole representative of the chordates, the phylum that gave rise to all vertebrate life.

chloroplast. Yet unknown processes are just as surely involved in the origin of the distinctive eukaryotic nucleus, mitosis and meiosis: the reductive division of chromosomes for the purpose of reproduction. The gulf between the prokaryote and the eukaryote is comfortably bridged, however, by the identity in all organisms of the genetic code, the universality of ATP as the energy-transferring molecule and the consistency of the amino acid composition of proteins in all forms of life. Such universals make sense only if they reflect a common ancestry.

What is the fossil evidence for the appearance of eukaryotes between two and 1.4 billion years ago? For one thing eukaryotic cells are generally larger than prokaryotic cells. As V. V. Timofeev and other Russian paleomicrobiologists have long noted, the average cell diameter of fossil microorganisms increases substantially in rocks of younger Proterozoic age. James W. Schopf of the University of California at Los Angeles has compiled data implying that the influx of larger cells took place about 1.4 billion years ago and involved a shift from sizes of generally less than 10 micrometers to commonly more than 20 micrometers. G. R. Licari and I have observed cell diameters up to 60 micrometers among microfossils of eastern California that may be 1.3 billion years old. Such dimensions imply a change then or earlier from an entirely prokaryotic microflora to a partly eukaryotic one. The geologic record for the interval between two and 1.4 billion years ago, however, is still sufficiently incomplete to leave open the possibility that older eukaryotes may yet come to light. It seems that oxygen levels would have favored their appearance at any time from about two billion years ago onward.

All in all there is little reasonable doubt that at some time between two and 1.4 billion years ago eukaryotic microorganisms, probably including green and red algae, became established. This opened the way to the biological precipitation of silica, which only eukaryotes can carry out (and which is probably why the silica in the banded iron formations seems to have been precipitated not biologically but chemically). It also satisfied one of the two essential preconditions for the evolution of the metazoans: the existence of the eukaryotic cell. The other precondition to be met, as Kenneth M. Towe of the Smithsonian Institution has pointed out, was a level of free oxygen high enough for the manufacture of such products as collagen.

After about 1.4 billion years ago the previously skimpy record of preserved cellular remains gets much better. It is as if evolution were speeding up. Yet it is another 700 million years before there is evidence of the oldest fossil organisms that all agree to be at once metazoan, indigenous to the sedimentary deposits in which they are found and deposited concurrently with those sediments. These organisms are the soft-bodied marine invertebrates of the Ediacarian System, which can be placed in the earliest part of the Paleozoic era and the Phanerozoic eon.

The Ediacarian fauna was originally known only from the Ediacara Hills of South Australia, but representatives of an even larger assemblage of fossils of this age have now been found in some two dozen different areas on five continents, entombed in sedimentary rocks deposited between 670 and 550 million years ago. Their appearance followed a nearly global succession of glacial deposits in later Proterozoic time. From such beginnings and from contemporaneous algal life there eventually arose the familiar biosphere of today.

The dominant organisms of the Ediacarian System, accounting for nearly 70 percent of the total, are coelenterates (more precisely Cnidaria), and of the coelenterates three-fourths are either jellyfishlike floaters or colonial forms reminiscent of the modern siphonophore *Velella,* drifters of the open sea and stranders on many a beach. The remaining coelenterates are colonial types that were attached to the sea floor like the modern sea pens. The noncoelenterates of the Ediacarian fauna include marine worms resembling modern polychaetes, some unusual arthropodlike animals that lack a carapace and a curious three-rayed discoidal animal that suggests a tiny naked starfish.

Some of these early drifters and inhabitants of the shallow sea floor left surface tracks and body imprints as records of the past, but none seems to have burrowed vertically into the bottom. Some were also quite large: jellyfish up to a meter in diameter and sea pens more than a meter long. A sheetlike marine worm called *Dickinsonia* grew to a length of nearly a meter but was less than three millimeters thick.

Although the Ediacarian animals are primitive, they are scarcely the kind of near-microscopic fauna that many paleontologists have expected to find at the base of the metazoan lineage. Still, in retrospect these strange animals are not unlike what one might expect if the reaching of some critical level of dissolved oxygen had triggered their evolution. As Rudolf A. Raff and Elizabeth C. Raff of Indiana University have demonstrated, jellyfish and similar coelenterates can get their oxygen through surface absorption at concentrations equivalent to only about 7 percent of the present atmospheric oxygen level. Under such conditions a thin, metabolically active surface combined with a large oxygen-collecting area would be advantageous. Thus it should not be surprising that jellyfish, thin-bodied worms and naked arthropods should predominate in the oldest known animal life. There is evidence that *Dickinsonia* had a gut and a weakly muscular body, implying that it also had an internal oxygen-collecting system. The Australian paleontologist Bruce Runnegar has calculated that even this form could have acquired enough oxygen at dissolved-oxygen pressures as low as from 6 to 10 percent of those near the sea surface today.

The fact that eukaryotic cells had apparently been around for 700 million years or more before the oldest known metazoans came along suggests that the triggering event for metazoan evolution was the cumulative buildup of oxygen in the hydrosphere and the atmosphere to somewhere near 7 percent of its present level. Quite possibly such a triggering was abetted by the ecological stress and geographic isolation arising from the drift of lithospheric plates, which at the time was rapid, and the related climatic changes. The first animals with phosphatic and calcareous external skeletons are found among the oldest Cambrian fossils and in rare instances among the youngest Ediacarian ones. Such impervious coverings would exclude the surface absorption of oxygen. By then gills and circulatory systems of some kind had presumably begun to function, suggesting an oxygen level perhaps closer to 10 percent of the present one. The race toward the more efficient utilization of biological energy was on.

Feedback effects related to the photosynthetic buildup of oxygen were far-reaching. In addition to its stimulation of eukaryotic and metazoan evolution, oxygen increased at the expense of carbon dioxide in both the hydrosphere and the atmosphere. The higher early carbon dioxide pressures probably account for the prevalence of dolomite over limestone among marine sediments during much of pre-Phanerozoic time. Nitrogen, which is the dominant atmospheric gas today, probably became so only gradually. Nevertheless, it was presumably always important. In the distant past it was a vital biospheric nutrient. It remains one today and is perhaps equally important as a nearly inert dilutant for corrosive oxygen.

The rest of biospheric history reflects the response of eukaryotic evolution to plate tectonics, climate and ecological challenge on the evolving Phanerozoic earth. It is a matter of detail, much of it beautiful and rich in significance. The Phanerozoic biosphere, descended from an almost entirely microbial one, has had a long history of ever more elaborate diversification that is still in progress. Where will it end? End it must, if only when the sun becomes a red-giant star four or five billion years from now.

THE AUTHORS
BIBLIOGRAPHIES
INDEX

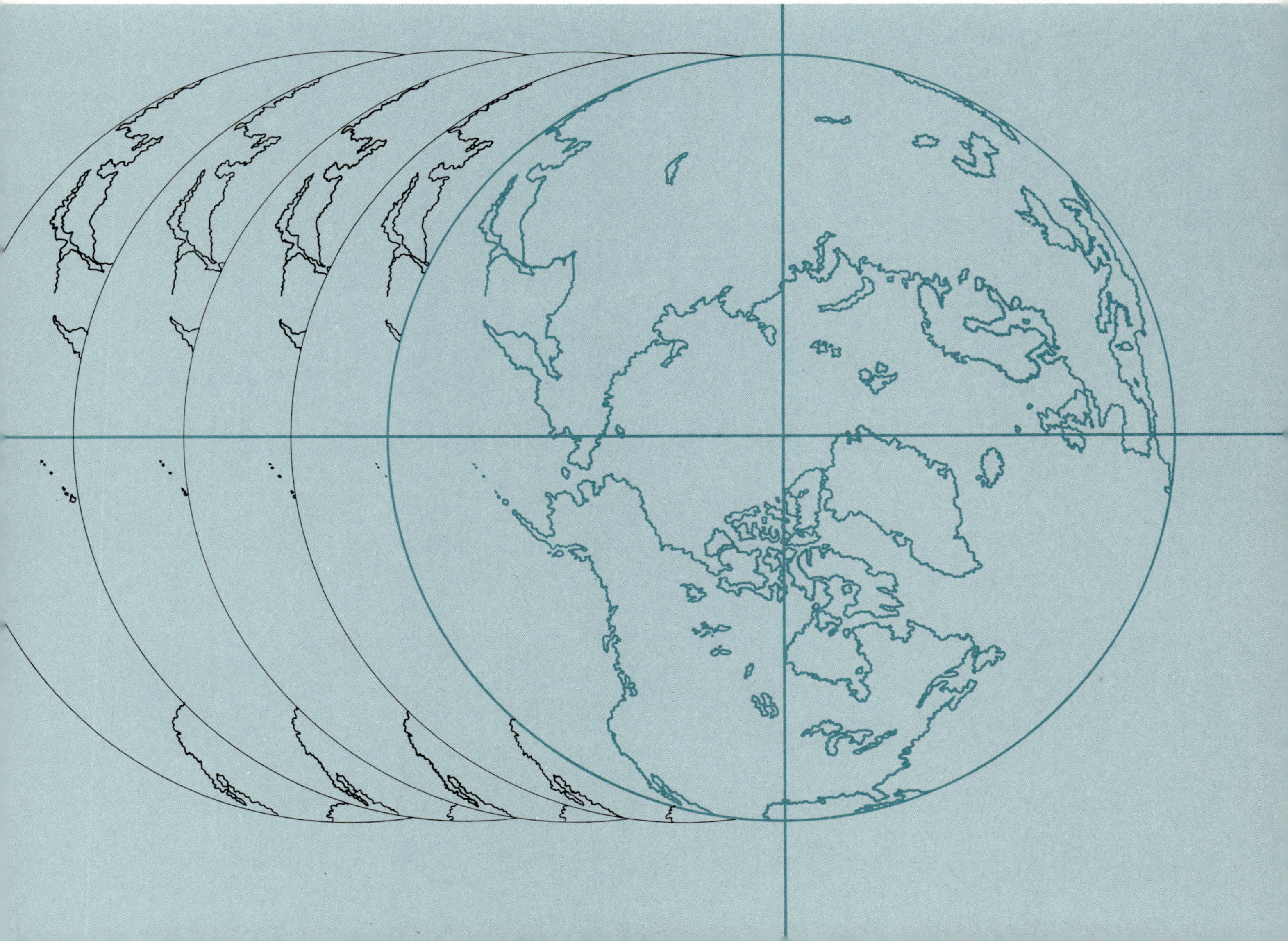

THE AUTHORS

RAYMOND SIEVER ("The Dynamic Earth") is professor of geology at Harvard University. He has three degrees from the University of Chicago: a B.S. (1943), an M.S. (1947) and a Ph.D. (1950). While he was still a graduate student he joined the Illinois Geological Survey, for which he worked until 1957. In 1956 he became a member of the Harvard faculty, and he has remained at Harvard since. He has served two stints as chairman of the department of geology. He recently stepped down for the second time to devote himself to teaching and experimental work. Siever writes: "My current interests are in the extremely low rates of geochemical reactions in buried sediments. I am fascinated by the relation of these processes to the internal heat of the earth, that is, how the formation of sedimentary rocks near the surface is affected by the dynamics of the interior, hundreds of kilometers below. Away from the university, the piano remains my abiding interest: alone, in duets or in chamber music."

RAYMOND JEANLOZ ("The Earth's Core") is associate professor of geology and geophysics at the University of California at Berkeley. He was graduated from Amherst College with a B.A. in 1975 and went on to obtain his Ph.D. in geology and geophysics at the California Institute of Technology in 1979. Jeanloz spent two years as a member of the faculty at Harvard University before moving to Berkeley in 1981. He writes: "My main interest is in mineral physics. I am particularly interested in studying how the physical and chemical properties of minerals control the ways in which planets evolve. Much of the work involves studying materials at the high pressures and temperatures that occur deep inside the planets. Like many earth scientists I became interested in geology through my enjoyment of the outdoors, in particular the high mountains."

D. P. McKENZIE ("The Earth's Mantle") is reader in tectonics at the University of Cambridge. His association with Cambridge goes back to his student days. He became a graduate student of Sir Edward Bullard's in 1963 and got his Ph.D. in geophysics at Cambridge in 1966. Soon after receiving his degree he joined the Cambridge faculty. He was promoted to his present job in 1979. He has served as visiting professor at several American institutions, most recently the University of Chicago. Most of McKenzie's scientific work has been on plate tectonics; he is currently most interested in the process of convection in the mantle and in the evolution of sedimentary basins.

JEAN FRANCHETEAU ("The Oceanic Crust") is a physicist on the staff of the Institut de Physique du Globe of the University of Paris. He is a native of France who received his diploma in mining engineering at the École des Mines in Nancy. He came to the U.S. to continue his education; the Scripps Institution of Oceanography awarded him a Ph.D. in geophysics in 1970. After obtaining his Ph.D. he returned to France to join the staff of the Centre Océanologique de Bretagne in Brest. He left the center in 1981 to go to the University of Paris. Francheteau's main scientific interest is the exploration of the ocean floor. His work has concerned the structure and morphology of the oceanic crust and the tectonics and geophysics of the ocean bottom. Much of his recent work has been done on the sea floor in the French submersible *Cyana*. He was the leader of Project RITA in 1978. During the expedition the first dives in a manned submersible to the crest of the East Pacific Rise were made in the *Cyana*.

B. CLARK BURCHFIEL ("The Continental Crust") is professor of geology at the Massachusetts Institute of Technology. His B.S., earned in 1957, is from Stanford University. He went on to get an M.S. at Stanford in 1958 before getting his Ph.D. in geology from Yale University in 1961. In the same year he joined the faculty at Rice University; he was at Rice until 1976. At the end of that time he moved to M.I.T. Burchfiel writes: "My current interest is in the process of orogenesis (mountain building) and its relation to the interactions of tectonic plates and to intraplate deformation. The specific areas I study are western North America, the Alpine mountain chains of eastern Europe and (more recently) of north-central China."

WALLACE S. BROECKER ("The Ocean") is professor of geology and director of the laboratory of geochemistry at Columbia University. His association with Columbia is more than three decades old. He received three academic degrees there: a B.A. in 1953, an M.A. in 1956 and a Ph.D. in 1958. In 1956, while he was still working on his Ph.D., he became a member of the Columbia faculty, and he has remained there since. Broecker's scientific concerns are diverse. He has done work on ancient climates and the causes of the cycles of glaciation on the earth, the large-scale mixing of water in the oceans and environmental geochemistry, including the effects of acid rain, metals in lakes and ocean chemistry.

ANDREW P. INGERSOLL ("The Atmosphere") is professor of planetary science at the California Institute of Technology. He attended Amherst College as an undergraduate, getting his B.A. in 1960. He received his postgraduate education at Harvard University, which awarded him an M.A. in 1961 and a Ph.D. in atmospheric physics in 1965. He writes: "My career goals took shape while I was an undergraduate physics major and a summer research assistant at the Woods Hole Oceanographic Institution. Since getting my Ph.D. I have devoted my professional life to the study of atmospheres and climates. I have participated in several National Aeronautics and Space Administration missions to the other planets, including the Pioneer missions to Jupiter and Saturn, the Pioneer mission to Venus and the Voyager mission, which flew by Jupiter and Saturn in 1979–81 and will fly by Uranus in 1986. I am also a member of a team that is analyzing the earth's radiation budget by means of data from a Nimbus spacecraft."

PRESTON CLOUD ("The Biosphere") is professor emeritus of biogeology and environmental studies at the University of California at Santa Barbara. He got a B.S. at George Washington University in 1938 and a Ph.D. from Yale University in 1940. From 1942 to 1961 he was on the staff of the U.S. Geological Survey. During his years with the Survey he held several different positions, among them that of chief of paleontology and stratigraphy. When he left, he became a member of the faculty at the University of Minnesota and the University of California at Los Angeles before moving to Santa Barbara in 1968. Cloud's main interest is the interaction of the physical environment and life in early times. He writes: "My interest in the origin of multicellular animal life, its antecedents and its first appearance on the earth has long been the central aspect of my research. In 1948 I proposed that multicellular animal life first appeared and evolved into a diversity of forms during an interval of less than 100 million years at the beginning of Phanerozoic time, a view that is now widely accepted. Using stromatolites, sedimentary structures, microbial fossils and the methods of ultramicroscopy and geochemistry, I traced the record of microbial life far behind the dawn of the Cambrian and integrated the study of the earliest biospheric, atmospheric and chemospheric evolutions and their interactions on the primitive earth."

BIBLIOGRAPHIES

Readers interested in further explanation of the subjects covered by the articles in this issue may find the following lists of publications helpful.

THE DYNAMIC EARTH

A REVOLUTION IN THE EARTH SCIENCES: FROM CONTINENTAL DRIFT TO PLATE TECTONICS. A. Hallam. Oxford University Press, 1973.

THE STEADY STATE OF THE EARTH'S CRUST, ATMOSPHERE AND OCEANS. Raymond Siever in *Scientific American,* Vol. 230, No. 6, pages 72–79; June, 1974.

ICE AGES: SOLVING THE MYSTERY. John Imbrie and Katherine Palmer Imbrie. Enslow Publishers, 1979.

LANGUAGE OF THE EARTH. Edited by Frank H. T. Rhodes and Richard O. Siever. Pergamon Press, 1981.

EARTH. Frank Press and Raymond Siever. W. H. Freeman and Company, 1982.

THE EARTH'S CORE

THE CURRENT STATUS OF SPECULATIONS ON THE COMPOSITION OF THE CORE OF THE EARTH. R. Brett in *Reviews of Geophysics and Space Physics,* Vol. 14, pages 375–383; 1976.

ORIGIN OF THE EARTH AND MOON. A. E. Ringwood. Springer-Verlag, 1978.

DRIVING MECHANISMS FOR THE EARTH'S DYNAMO. D. Gubbins and T. G. Masters in *Advances in Geophysics,* Vol. 21, pages 1–50; 1979.

BASALTIC VOLCANISM ON THE TERRESTRIAL PLANETS. Basaltic Volcanism Study Project. Pergamon Press, 1981.

MODELS OF THE EARTH'S CORE. D. J. Stevenson in *Science,* Vol. 214, No. 4521, pages 611–619; November 6, 1981.

THE INTERIOR OF THE EARTH: ITS STRUCTURE, CONSTITUTION AND EVOLUTION. M. H. P. Bott. Elsevier Science Publishing Co., Inc., 1982.

RECENT DEVELOPMENTS IN THE DYNAMIC THEORY OF PLANETARY MAGNETISM. F. H. Busse in *Annual Review of Earth and Planetary Sciences,* Vol. 11, pages 241–268; 1983.

THE EARTH'S MANTLE

CONVECTION IN THE EARTH'S MANTLE: TOWARDS A NUMERICAL SIMULATION. D. P. McKenzie, J. M. Roberts and N. O. Weiss in *Journal of Fluid Mechanics,* Vol. 62, Part 3, pages 465–538; February 11, 1974.

THE CHEMICAL ORIGIN OF THE EARTH'S MANTLE. R. K. O'Nions, P. J. Hamilton and Norman M. Evenson in *Scientific American,* Vol. 242, No. 5, pages 120–133; May, 1980.

PLANFORM OF MANTLE CONVECTION BENEATH THE PACIFIC OCEAN. Dan P. McKenzie, Anthony Watts, Barry Parsons and Micheline Roufosse in *Nature,* Vol. 288, No. 5790, pages 442–446; December 4, 1980.

THE OCEANIC CRUST

THE FACE OF THE DEEP. Bruce C. Heezen and Charles D. Hollister. Oxford University Press, 1971.

PLATE TECTONICS. Xavier Le Pichon and Jean Francheteau. Elsevier-Scientific Publishing Company, 1973.

PHOTOGRAPHIC ATLAS OF THE MID-ATLANTIC RIDGE RIFT VALLEY. Robert D. Ballard and James G. Moore. Springer-Verlag, 1977.

BIRTH OF AN OCEAN: THE CREST OF THE EAST PACIFIC RISE. Jean Francheteau, Thierry Juteau, David Needham and Claude Rangin. Centre National pour l'Exploitation des Océans, 1980.

THE CONTINENTAL CRUST

THE EVOLVING CONTINENTS. Brian F. Windley. John Wiley & Sons, Inc., 1977.

EVOLUTION OF THE EARTH'S CRUST. Edited by D. H. Tarling. Academic Press, 1978.

THE CONTINENTAL CRUST AND ITS MINERAL DEPOSITS. David W. Strangway. Special Paper No. 20, Geological Society of Canada, 1980.

STUDIES IN GEOPHYSICS: CONTINENTAL TECTONICS. Edited by B. C. Burchfiel, J. E. Oliver and L. T. Silver. National Academy of Sciences, 1980.

THE GROWTH OF WESTERN NORTH AMERICA. David L. Jones, Allan Cox, Peter Coney and Myrl Beck in *Scientific American,* Vol. 247, No. 5, pages 70–84; November, 1982.

THE OCEAN

TROPICAL RAINFOREST HISTORY AND THE EQUATORIAL PACIFIC CARBONATE DISSOLUTION CYCLES. N. J. Shackleton in *The Fate of Fossil Fuel CO_2 in the Oceans,* edited by Neil R. Andersen and Alexander Malahoff. Plenum Press, 1977.

DEEP-SEA CARBONATE AND THE DEGLACIATION PRESERVATION SPIKE IN PTEROPODS AND FORAMINIFERA. W. H. Berger in *Nature,* Vol. 269, No. 5626, pages 301–303; September 22, 1977.

GAIA: A NEW LOOK AT LIFE ON EARTH. J. E. Lovelock. Oxford University Press, 1979.

EXTRATERRESTRIAL CAUSE FOR THE CRETACEOUS-TERTIARY EXTINCTION. Luis W. Alvarez, Walter Alvarez, Frank Asaro and Helen V. Michel in *Science,* Vol. 208, No. 4448, pages 1095–1108; June 6, 1980.

TRACERS IN THE SEA. Wallace S. Broecker and T.-H. Peng. Lamont-Doherty Geological Observatory, Columbia University. Eldizia Press, Palisades, New York, 1982.

ICE CORE SAMPLE MEASUREMENTS GIVE ATMOSPHERIC CO_2 CONTENT DURING THE PAST 40,000 YR. A. Neftel, H. Oeschger, J. Schwander, B. Stauffer and R. Zumbrunn in *Nature,* Vol. 295, No. 5846, pages 220–223; January 21, 1982.

THE ATMOSPHERE

VARIATIONS IN THE EARTH'S ORBIT: PACEMAKER OF THE ICE AGES. J. D. Hays, John Imbrie and N. J. Shackleton in *Science,* Vol. 194, No. 4270, pages 1121–1132; December 10, 1976.

CLIMATIC CHANGE ON THE TERRESTRIAL PLANETS. James B. Pollack in *Icarus,* Vol. 37, No. 3, pages 479–553; March, 1979.

CLIMATE IN EARTH HISTORY. Wolfgang H. Berger and J. C. Crowell. National Academy Press, 1982.

THE BIOSPHERE

THE BIOSPHERE. G. Evelyn Hutchinson in *Scientific American,* Vol. 223, No. 3, pages 44–53; September, 1970.

COSMOS, EARTH, AND MAN: A SHORT HISTORY OF THE UNIVERSE. Preston Cloud. Yale University Press, 1978.

SYMBIOSIS IN CELL EVOLUTION: LIFE AND ITS ENVIRONMENT ON THE EARLY EARTH. Lynn Margulis. W. H. Freeman and Company, 1981.

THE EDIACARIAN PERIOD AND SYSTEM: METAZOA INHERIT THE EARTH. Preston Cloud and Martin F. Glaessner in *Science,* Vol. 217, No. 4562, pages 783–792; August 27, 1982.

INDEX

References to illustrations appear in boldface type.